Systems Engineering Neural Networks

Systems Engineering Neural Networks

Alessandro Migliaccio
Giovanni Iannone

Registered Office
John Wiley & Sons, Inc., 111 River Street, Hoboken, NJ 07030, USA

For details of our global editorial offices, customer services, and more information about Wiley products visit us at www.wiley.com.

Wiley also publishes its books in a variety of electronic formats and by print-on-demand. Some content that appears in standard print versions of this book may not be available in other formats.

Library of Congress Cataloging-in-Publication Data Applied for:

ISBN - 9781119901990 (hardback)

Cover Design: Wiley
Cover Image: © Peera_stockfoto/Shutterstock

Set in 9.5/12.5pt STIXTwoText by Straive, Chennai, India

... look at the world through your own eyes and always be self-aware.

Contents

About the Authors

ALESSANDRO MIGLIACCIO, CEng, ASEP: graduated in Space Systems Engineering at Delft University of Technology, currently working as a Systems Engineering Development Leader at Airbus. Expert in mixed reality technology and aspiring futurist, with more than 10 years of work experience in aeronautical companies in several countries, he has led data analytics projects aimed at the optimization and tailoring of maintenance programs of airplanes. A Chartered Engineer from the Royal Aeronautical Society and certified systems engineering practitioner, he is passionate about refining his skills by finding new ways to improve team work with new paradigms based on professional fulfillment, nurturing of talents and democratization of niche technologies. To this aim he founded AIShed as an association of voluntary talents dedicated to outreach and research. LEGO builder and drone pilot, he enjoys volunteering work as STEM ambassador and practicing mindfulness.

GIOVANNI IANNONE: graduated in Mechanical Engineering for Design and Manufacturing at "Università degli Studi di Napoli," and was awarded a Master in Systems Engineering at MS&T (U.S.). Expert in aeronautical structures and continuous airworthiness on large airplanes, with more than 10 years of work experience in Subject Matter Expert (SME) departments. He has taken a critical interest in the decision making process by different mathematical approaches. Member of INCOSE for several years, he presented at ASEC2014 on decision making algorithms in the sports business.

Acknowledgements

This written work is the result of a year's work and the product of a decade-long professional partnership based on a continuous exchange of ideas. As it is often the case, very few pieces of work come from individual minds. These usually grow in a (neural) network of friends and supporters, through late night conversations and amiable chats by the coffee machine. Eventually the activity has become so large as to require structure, therefore we launched AiShed (Figure 1), a cultural association headquartered in France, devoted to outreach and to fuel a community of likeminded people with an interest in AI.

Firstly, we must thank our families, friends and colleagues - too many are their names for a space this small. A special thank you goes to Federica Migliaccio and Giuliano Iannone for their relentless support of the AiShed association activities and to Angela Masella for the excellent work of translation and editing. A note of gratitude goes to Eberhard Gill, professor of Space Systems Engineering and Director TU Delft Space Institute for introducing Systems Engineering to bright young minds every year, including my own. Furthermore, a dear thank you to Klein Kim for her work on the athletic performance code and Daniel West for allowing us to share his curious and interesting work on the LEGO® sorting machines.

Thank you to all of you.

Figure 1 www.ai-shed.com.

How to Read this Book

This book is not meant to compete with all the research that made it possible for more and more innovative machines to be developed. Our aim is to simply help the reader put into practice the theories presented in the next pages through the useful applications found in the last part of the text. The content of the book revolves around the lifecycle of a generic system (Figure 1) and addresses how Neural Networks can be called into actions in different phases of a system development.

The text is structured in three parts with the first part focused on systems engineering while the second will present a number of exercises through the use of two programming languages, Python and Visual Basic. This will allow readers of different academic backgrounds to interact with neural networks. Part 3 covers the theory of Neural Networks and its key components.

Chapter 3 is particularly interesting, as we will delve further into the link between the theory of Systems Engineering and Analytic Foresight. An innovative approach will be used to show how to apply neural networks to the sports business. A quirky example is the one related to LEGO® sorting machines - as unusual as it may sound, sorting machines are at the basis of industrial engineering, from automotive applications to food technology.

The journey to understanding neural networks is a fascinating one though it can be perceived as arduous to the inexperienced reader. This topic requires an academic knowledge of basic calculus. Let us reach an agreement: in this book we will examine the basics of the topic, assuming that the reader will be proactive in utilizing the resources available on the Internet or in literature to close gaps in understanding.

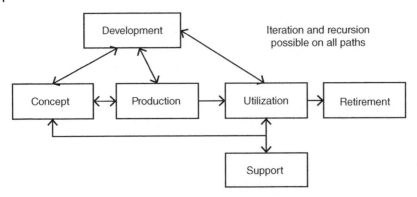

Figure 1 Life cycle model with some of the possible progressions. Source: INCOSE Systems Engineering Handbook: A Guide for System Life Cycle Processes and Activities, Forth edition– Wiley.

You can read the book chapter by chapter or by picking a specific chapter of interest. This book is enriched with examples from classic literature and examples from industry to help you grasp the most difficult notions. There is a glossary in "Glossary and Insights" to include other clarifications and references to in-depth studies.

We have also created a website we invite you to visit, hoping to intrigue your curiosity: http://www.ai-shed.com.

We hope you find this an entertaining and useful read!

Part I

Setting the Scene

1

A Brief Introduction

I see it all perfectly; there are two possible situations – one can either do this or that. My honest opinion and my friendly advice is this: do it or do not do it – you will regret both.

Søren Kierkegaard

From the Ancient Greeks through the Renaissance, and until our present day, human beings have always tried to give meaning to the reality surrounding them. This effort was not based on tradition or myth, but on the human rational ability to describe reality through the laws of mathematics.

When writing a scientific text and trying to give reality a meaning by applying a mathematical model, we cannot ignore certain philosophical concepts. On the contrary, we must find inspiration in the opinions of the great thinkers of the past. We will only examine a few postulates, but you can rest assured that many more are available and extensively explained in the literature. We take advantage of the knowledge that was made available to us by such human talents.

Our calculus teachers would never stop saying that numbers have to be interpreted, understood, and explained organically. Thanks to numbers we can define an object, an event[26], or a physical phenomenon. Why is there a need to interpret numbers?

According to Pythagoras, numbers are the primordial elements from which reality is derived. The latter can be inferred through a strict mathematical and geometric sequence. The qualitative and contemplative elements coexist with the quantitative one. Each number is associated with a shape containing elements which allows them to stay together in a harmonized and neat manner. Therefore, if we base our interpretation of the world on its numerical and harmonious nature, we can come to understand it starting from its measurements. Are numbers all we need to understand the world? What is your idea of the reality surrounding us?

It is not easy to have an idea and expand upon it – we could find it difficult, for example, to distinguish true from false and zero from one. Once its traits are

Systems Engineering Neural Networks, First Edition. Alessandro Migliaccio and Giovanni Iannone.
© 2023 John Wiley & Sons, Inc. Published 2023 by John Wiley & Sons, Inc.

defined – zero or one, true or false – an idea is absolute and unalterable, therefore we can associate it to reality.

To paraphrase the words of Plato and inferring his theories from his dialogs – we apologize in advance to our fellow philosophers and teachers of philosophy – the Idea exists outside of our mind. It is detectable only by our intellect and not by our perception, the latter being not sufficient to understand reality. We can simplify by saying that the idea is similar to a standard of judgment. We have access to the knowledge of things only if we have ideas (See Figure 1.1).

Ideas act as measurements to evaluate the tangible reality, and they do not reside in our imagination. The idea, as an objective entity, is not to be confused with opinion, which is instead subjective. As we know in physics and mathematics, we have to measure the phenomena we observe daily with certainty and precision. The idea is a model (or archetype) correlated to our empirical world – we should only try to imitate or duplicate this model. We will see later on how this model acts as an absolute reference for our implementation.

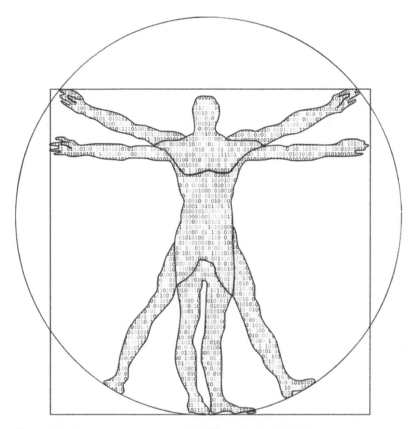

Figure 1.1 The human being becomes the standard for all things.

If we wanted to make a measurement of a particular event and then attribute to that event a meaning, we could start with the concept of an idea, in an absolute sense, as an essential reference. On the other hand, if the measurement is associated with a judgment, then we cannot know whether there is an absolute rule to discriminate that event. Surely we can trace back through experience to the general rule governing an event.

Therefore, as we have mentioned earlier in this chapter, human beings have always felt the need to explain the worldly reality they live in and, might we add, it could not be any other way. Man's senses can be accepted as an important source of knowledge.

However, would we as human beings be able to understand the world based on rules established by our rationality? The Vitruvian Man, famous work by Leonardo da Vinci, conveys a model of a human body that is analyzed and measured through mathematical and geometric tools. The human being becomes the standard for all things, therefore humanity as a whole gains full awareness. Man2, put at the center of the world, becomes the symbol of a better future.

We now have all the elements to start writing about mathematical models, which can also be referred to as rational and variable structures, integrated in logical processes. These structures are based on the concepts of Number, Idea and Human, which we have introduced earlier.

The ultimate meaning of an event is seen as a reliable reference, and we aim toward it. A systematic and methodical approach to the analysis of an event – as we will see in Chapter 3 – will help us interpret it. Its modeling can take us to more reliable, though not absolute, conclusions.

When applied to varied events in our lives, the use of mathematical models can help the reader to better interpret certain dynamics that are part of our daily life. The examples in this book are relevant to those aspects that are often difficult to decipher due to their complex nature. Processes such as decisional ones can be understood via computational models used in the examples provided.

The German physicist W.K. Heisenberg affirmed that concepts of probability apply to all cognitive processes. Human beings would not be able to reach a perfect understanding of a physical phenomenon because the observer is not able to determine how they are interfering with the observed object.

To build a mathematical model we need to define a prerequisite that establishes its efficacy: this is uncertainty (Figure 1.2). Uncertainty obviously plays a vital role in the development of complex models, as they require a reliable mathematical form to describe a real problem. We would be ignoring our reality, and our understanding of it, if we ignored the element of uncertainty. It is easily understandable that closed-form expressions offer a description of simple and less uncertain realities, which is more accurate.

But what if reality is instead complex, uncertain, and difficult to interpret?

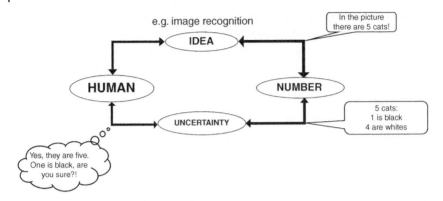

Figure 1.2 Number, Idea, Human, Uncertainty…what else?

Now, this is our objective: let us develop a tool that operates through estimates and aims at minimizing the possibility of error so that we can rely on a result as close to objectivity as possible.

How can we achieve that?

Below are two images, at first glance remarkably similar, depicting two different animals: a cheetah and a leopard (Figure 1.3). Would you be able to tell them apart and say which is which? How can we define the two animals if we do not have an extensive knowledge of zoology? The obvious solution, not to be dismissed from the start, would be to ask an expert and have him explain to us how to tell them apart. We can obtain an accurate solution to the problem if we make use of specific knowledge to define a series of characteristics. The solution, as a result of data processing, will be more accurate if we can count on all the applicable variables of the problem and if we can relate these to the characteristics of the animal depicted below.

Figure 1.3 Cheetah or Leopard? Could you tell them apart? Source: Image by Jonathan Reichel from Pixabay.

The approach we take in this book is a common application of linear and non-linear algebraic combinations – these somehow describe the various interactions happening in our brain when it is prompted to find a solution to a problem. The individual outcome of these efforts can be defined as the union of elements and variables working together to perform a specific function.

The reader will find a brief description of neural networks, as a calculation methodology, and some cases, taking the opportunity at the same time to exemplify the "system approach" used throughout the book. It is important to clarify that the learning phase is essential for the neural network to work as expected. A network can detect the trends in the data regardless of how they fluctuate, based on the exact behavior of all the involved variables. The scope of machine learning[3] as a discipline is to find a correlation between historic data and present (and future) data; when the correlation is found, the network detects it and uses it as the basis of its prediction[32].

Even a distracted reader can see that predictions are valid only if future data trends align with past ones. In other words, if we think of an unchangeable, static and frozen reality, we could set in stone all the algorithms that work successfully. Our reality is constantly changing, but we have to start somewhere – won't you agree? Therefore, let us build the basis of our models and then create an algorithm that can guarantee a certain degree of accuracy in analyzing specific dynamics, and also reduce uncertainty[34] as much as possible. In fact, the continuous changes of events can be managed by adapting and improving the algorithms.

Let us take the human brain as a reference to explain this better – after all, we have always been told that our brain is the biggest computer ever created.

Pieces of information move inside a neural circuit of two or more neurons, thanks to a communication process based on chemicals and electrical impulses that is repeated billions of times. Any piece of information in our brain travels at extremely high speed until it reaches the cerebral cortex where information is analyzed and ultimately "understood."

Before reaching the cortex, the information will go through a high number of synapses. As the same process is repeated time and time again, information will follow the same well-known route. Let us recap: as pieces of information go through the same process a number of times, synapses become so used to it that even different, but relevant, signals ignite the same sequence of impulses. The entire process of memorization is part of a "simplified" process that, starting always from the same point, travels easily through the same steps again and again.

The neural networks' learning process, as presented in this book, follows the general principles described above, though the process happening in our brain is way more complex than the one we will use for our artificial neural networks. Our brain is constantly reshaping, and old synapses are replaced by new ones when our brain goes through learning or memorization phases. On the contrary, the networks used in this book will remain quasi-static[46] once they are set up and will be used to process the same type of information.

We hope that the notion of "learning" is now established as the basis of the implementation of all neural network algorithms. Once all scenarios and conditions are memorized, we can proceed and estimate ways to learn the difference between right and wrong. Moreover, it is possible to only learn from pieces of information that are not decoded or limit our research to the correlation present among the available data. It is also possible to learn from known data and make predictions or decisions by comparing all the available options once the data is processed. If we think that reality is way more complex, the choice of learning and processing data with linear models might be oversimplifying. Our aim is to uncover the correlation that exists among the known data, or rather the physics and mathematical models at the basis of all the possible scenarios that the network needs to crunch.

We have noticed how cognitive science and neuroscience have come closer to each other in the last millennium. The technological advancement and research made by physicists, psychologists, and neurophysiology experts on the study of the brain enable us to develop more complex neural networks (Figure 1.4).

We would like the reader to focus on the concept of learning once more, and to try and optimize it so as to distinguish it from other concepts or processes. How can we improve the learning phase? Can we just increase the number of data or improve the quality of the data itself? By adding structure to the network, we would be increasing the complexity of the code and the processing time.

Figure 1.4 How many scenarios of a GO game can you imagine? Source: Image by Jonathan Reichel from Pixabay.

It is known that when we focus too much on the details of a simple problem, we lose sight of the context, and we can easily find ourselves with a foggy solution. Equally, if we had access to high processing capability but insufficient training data, we would need to steer the learning process toward a specific solution, rather than a generic one.

It is safe to affirm that a mathematical structure similar to the human brain cannot replace the reasoning of every human being, especially when it comes to the ability of a human being to make decisions based on their feelings or intuition – we know how these elements can at times be more efficient than reason! From a mathematical point of view, our aim is to help the reader to process the largest share of data that is possible and reach a reliable conclusion from a mathematical point of view. If you think how difficult it is to create a model based on a worm's 300 neurons, you can easily understand how arduous or impossible it would be to reproduce a model of the human brain, with all its 85 billion neurons. This is not surprising if we consider that the human brain has developed throughout millions of years and AI is a subject that was only born in the last century.

Note

In the book we shall refer to Neural Networks and Deep Learning[4] as AI for simplicity, keeping in mind that these are only some of the techniques covered in the wide spectrum of AI technology.

1.1 The Systems Engineering Approach to Artificial Intelligence (AI)

Systems Engineering (SE) represents an important paradigm to support the introduction of new process activities and new technology. The systems engineering approach emerged as an effective way to manage complexity and change and supplies a consistent framework to engineers who are willing to design intelligent systems. Chapter 3 will show how ultimately it takes no additional systems engineering effort to design and manage systems heavily reliant on AI. Nevertheless, it would be blind to ignore the necessity of injecting AI capabilities to augment the performance of existing systems. To put in simpler word, it is the opinion of the authors that when talking about systems making use of artificial intelligence (AI), systems engineering becomes, in essence, "common sense engineering."

From this point on we shall refer to ANN (Artificial Neural Network) and AI subsystems to indicate a black-box module performing machine learning functions. It is a good idea to start this endeavor with a clear understanding the existing system and look for points in common with those of an ANN-based system, from this point on called simply AI. An interesting starting point is that building a system around AI is not the most common approach. AI is most often a supporting

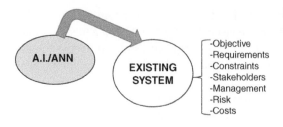

Figure 1.5 Integration of AI/ANN into an existing system.

element and can be a part of the existing system, usually with the aim of improving system performance such as its accuracy or reliability. Most of the examples presented in this book, from a spacecraft (system) or autopilot (subsystem), to the corrosion detection are existing systems that can be augmented with AI, while others like the LEGO® Sorting Machine is a system designed around the use of AI to perform its core function.

How the AI can be integrated into an existing system? Starting from system architecture is the best practice to have the correct knowledge of the system and its components (Figure 1.5).

When it comes to neural networks, the highest level of detail is certainly coded by a defined programming language. The entire system architecture should be clear to the system engineer and allow to readily extrapolate the part of the system in which the ANN is to be introduced. The now called AI-subsystem itself and the interfaces with other systems shall provide the available dataset, needed to build the algorithm. The in-depth knowledge of each system function will lead to defining the objective of the AI-subsystem and above all, a rigorous management of the constraints and requirements. In any field of work, the techniques for managing AI within a complex system are always the same and the reader should acquire the critical skills needed to make the best use of the systems engineering approach.

The first steps in the systems engineering approach to solve any technological or non-technological challenge are connected to answering certain basic questions: Who needs the system? What is the system meant to accomplish? Who is going to operate it? Who will benefit from the successful operation of the system?

By answering these questions, the developer of such a system will be able to setup a rigorous set of requirements and constraints, before moving on with the design, prototyping, verification and validation activities mentioned in Chapter 3 and in most of the literature (see INCOSE Systems Engineering Handbook).

Some of the more pragmatic readers, already familiar with the principles of systems engineering may think that the many software packages used to manage this complexity would overwhelm the "common sense engineering" mentioned above, that is not necessary as "system thinking" is not in contradiction with pragmatism.

Developing the system architecture is one of the most important phases of the SE approach. Creating an effective architecture draws on the experience, intuition, and good judgment of the engineers to devise an appropriate solution. The systems architecture builds on four methodologies:

- Normative (solution based) such as building codes and communication standards.
- Rational (method based) such as systems analysis and engineering.
- Participative (stakeholder based) such as concurrent engineering and brainstorming.
- Heuristic (lessons learned) such as simplify the framework.

Why is architecture an important phase? Because systems architecture is an adequate process, and because intuition and experience play such an important role, the systems engineer must pay attention to situations where past experience and intuition have been a handicap.

Even after all new system elements are deployed, (Figure 1.6) product/project management must continue to account for changes in the various system element life cycles. A new technology will impact one or more system elements, and an existing system is superseded by such improvement.

How is the correct integration reached? At first, the AI inputs and outputs have to be compatible with the existing system or its components. Such compatibility of components of a large and complex system is needed to work as a single entity. This is an important feature of AI. It can work autonomously in a complex system and can continue to improve when the system changes.

The AI capability is essential to avoid redesigning the existing product/project which cannot be easily reworked due to delays with concomitant cost overruns. Eventually, a production engineering analysis for each design alternative becomes an integral part of the Architectural Design process. The production engineering creates any constraints on the existing design. Such constraints shall be communicated and documented. For complex systems, a multidisciplinary analysis is a key task to clarify the existing design and reduce risk, lead time, and cycle time;

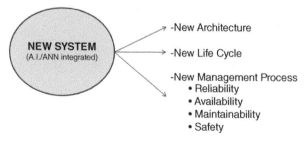

Figure 1.6 A new system is met when an AI is introduced.

and to minimize strategic or critical materials use. The design clarification should account for assembly and disassembly of the integrated AI for maintenance.

The existing system is surely supported by a qualified process. If the existing process remains satisfactory when the AI is integrated, then the AI impact on risks and cost-effectiveness becomes evident. The program risk analysis, if necessary, is a next phase to be performed to evaluate AI long-lead-time items, and special processes.

The next step is to share with the working team what is being done. In particular, it will be communicated as following:

- the level of uncertainty,
- the degree of complexity,
- the consequences to human welfare.

All systems work in an environment influenced by events (Figure 1.7). The events definition is the key to establishing the tasks and activities that need to be completed prior to introducing the AI. All projects are subject to uncertainty; an uncertain event may be harmful if it occurs (threats), another may assist in achieving objectives (opportunities). Dealing with both types of uncertainty under the single heading of "risk management" minimizes process and overhead and expands organizational and personal commitment toward finding and capturing opportunities. Since traditionally, project managers think of risks as threats alone, it may be a change to begin recognizing opportunities. If opportunities are handled along with threats, risk management language needs to be balanced

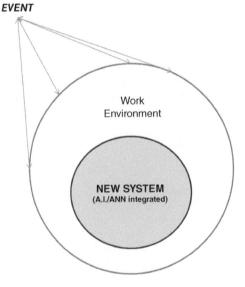

EVENT

Work Environment

NEW SYSTEM
(A.I./ANN integrated)

Figure 1.7 Development and deployment of a new system in the work environment.

with terms for opportunities such as "exploit," "share" and "protect." In this book the proper balance between the risk of missing project technical and business objectives on the one hand, and process paralysis on the other is not discussed.

Anyway, risk and opportunity management is a disciplined approach to dealing with uncertainty that is present throughout the entire system's life cycle. This process is used to understand and avoid the potential cost, schedule, and performance/technical risks to a system, and to take a proactive and structured approach to anticipate negative outcomes, respond to them if they occur; and to identify potential opportunities that may be hidden in the situation. Every new system or modification of an existing system is based on pursuit of an opportunity.

Risk always is present in the life cycle of systems, and the risk management actions are assessed in terms of the opportunity being pursued. Risk can also be introduced during architectural design caused by the internal interfaces that exist between the system elements. System development may be rushed to deploy the system as soon as possible to exploit a marketing opportunity or meet an imminent threat, leading to schedule risk. It will be stressed once more that a rigorous risk management process, in line with the project management doctrine as in SE will allow us to successfully integrate the new risks connected to the introduction of AI into the system, as well as the risks removed by such introduction.

In addition, it is important to mention the configuration control phase. Integrating AI in an existing system means making a change to a system. In this book AI is shown as a software referred to improve and control some product/process. The baselines are established by review and acceptance of requirements, design, and product specification documents. The creation of a baseline may coincide with a project milestone or decision gate. As the system matures and moves through the life cycle stages the AI is maintained under configuration control.

AI can be used to manage a lot of information and, as the system engineer, it would balance a complex system looking for optimal points between its elements, embedding in the system processes the concepts of AI, (AI for SE). On the contrary, the AI integration must not be concerned that a stand-alone product is used to solve a specific technical problem. For instance, considering that the airplane autopilot can be changed into an autonomous system, the next higher system, aircraft, is again considered in order to find the balance points and to create a new balanced system, (SE for AI). In general, we could theorize that AI must be embedded in a complex system, but it should not be seen as software since the right effectiveness cannot be achieved. In summary, any complex system that involves non-linear analysis of its processes to achieve a defined objective can be managed by the integration of AI/ANN.

By introducing the ANN as a Configuration Item (software/application) of a new system, this means that it is designated for a separate control of the New System Configuration. The Configuration Management (CM) shall ensure the control of the ANN functional, performance, and physical characteristics that are properly identified in the newly designed system. Hence, the CM helps to establish and maintain control of requirements, documentation, and artifacts produced throughout the system's life cycle. If the change is necessary, the AI impact shall be clarified, especially defining effects on the life cycle. This can increase risk which can adversely affect system cost, performance, and safety.

CM is the practice of applying technical and administrative direction, surveillance, and services to:

- Identify and document the characteristics of system elements such that they are unique and accessible in some form; assign a unique identifier to each version of each system element.
- Establish controls to allow changes in those characteristics; ensure consistent product versions.
- Record, track, and report status pertaining to change requests or problems with a product; maintain comprehensive traceability of all transactions.

1.2 Chapter Summary

Systems engineering is introduced here as a fertile ground for the application of neural networks. This chapter introduces the reader to the concepts that inspired the authors, and to some of the ideas presented in the book. It aims to put the reader in the right mind-set but also to give some useful guidance on the structure of the book. It also establishes a "pact" with the reader, aimed at a more effective use of the book. In the second part of the chapter the key connection between AI and Systems Thinking is established.

Take outs:

- Human beings have felt the need to explain the worldly reality they live in. However, we are able to understand the world by our rationality. The ultimate meaning of an event is seen as a reliable reference, and a systematic and methodical approach will help us interpret it.
- Our objective is to develop a mathematical tool that is more accurate when reality is complex, uncertain, and difficult to interpret. We apply linear and non-linear algebraic combinations to describe the various interactions happening in our brain when we solve problems.
- The human brain is the biggest computer ever created. The brain communicates by chemicals and electrical impulses, and the information is analyzed in the cerebral cortex.
- Neural networks are constantly reshaping their connections, but artificial neural networks remain quasi-static, so they cannot be used to analyze athlete's performance by using engine data.

- Systems Engineering is an important paradigm that supports the introduction of new technology and processes. AI can be integrated into existing systems using systems engineering.
- Starting from system architecture, in-depth knowledge of the system and its components, knowledge of the system's requirements and constraints, and knowledge of the AI-subsystem is the best practice for system integration.
- Risk and opportunity management is an approach to dealing with uncertainty throughout the entire life cycle of a system and anticipating negative outcomes.
- AI can be used to manage complex systems and to find optimal points between its elements, and must be embedded in the system processes to achieve the right effectiveness.

Questions

1   To extend the mathematics laws to reality, why do we need to consider uncertainties?

2   How can the SE approach support any Neural Networks application?

3   How can Neural Networks help manage complex systems?

4   Where can a Neural Network be introduced into an existing system?

5   When the system is already disposed, what processes will be used to introduce the Neural Network?

2

Defining a Neural Network

Recently, neural networks and deep learning have attracted even more attention with their successes being regularly reported by both the scientific and mainstream media, […] or the more recent AlphaStar. This renewed interest is partially due to the access to open-source libraries such as TensorFlow, PyTorch, Keras or Flux.jl to name just a few.

Jean-Christophe B. Loiseau

Have you ever heard of neural networks? What about AI? Are neural networks and AI synonyms? In this chapter, we will try to answer these questions by looking at the interesting comparison between AI and the human brain.

You could be thinking about some science fiction watched recently, starring a hero fighting against dozens of robots at the top of a building in flames. Reality is not too different from our imagination, even though we have to allow for exceptions. Can we consider artificial networks to be similar to biological networks?

It is usually considered inappropriate or even misleading to start with this comparison, even though there are some evident parallels that come to mind. Before we dive deep into these concepts, let us explore the landscape a bit.

We could say that neural networks are a subcategory of AI, the same as machine learning, deep learning and computer vision.

The concept of a network will be well known to individuals who are familiar with AI (Figure 2.1).

A network is a group of interconnected elements constantly exchanging information, exactly as the human nervous system which is made of a group of special cells connected by interacting nervous fibers. These cells are able to exchange numerous pieces of information.

The neural network is usually inspired by some aspects of biological neural networks. It is not surprising that in the last years we have seen a growing interest in these similarities, which could help develop parallel and distributed computing systems[14].

Systems Engineering Neural Networks, First Edition. Alessandro Migliaccio and Giovanni Iannone.
© 2023 John Wiley & Sons, Inc. Published 2023 by John Wiley & Sons, Inc.

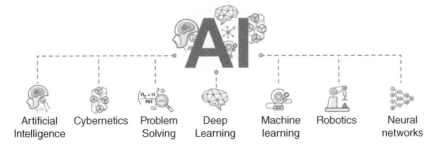

Artificial Intelligence Cybernetics Problem Solving Deep Learning Machine learning Robotics Neural networks

Figure 2.1 How many branches of knowledge are linked to AI? Here are some examples. Source: Shutterstock.

The neural networks described in this book are an immensely powerful tool for automatic processing and they deal with a big amount of data.

Recently we have observed a peculiar feature of a specific cat breed – Manx cats, originating in the Isle of Man, are genetically without a tail (Figure 2.2). This feature could compromise computational analysis as, if a tail was defined as a distinctive element to identify a cat, then the neural network would have struggled with such an identification task. To be accurate, the network could still correctly identify the cat depicted here, by processing a large amount of data and dismissing the importance of a piece of data over another.

Figure 2.2 Is it or is it not a cat? The Manx cat dilemma. Source: Image by spicetree687 from Pixabay.

This example shows how a human being would easily identify the animal as a cat, dismissing the detail concerning its tail. This is different from what we would see in AI, as a specific training would be needed to reach such identification easily.

2.1 Biological Networks

The human brain is made of 86 billion nerve cells, called neurons. Neurons are special cells which can produce and transmit electrical signals.

Let us go back to the discovery of neurons. Neurons were discovered in 1879, when a laboratory assistant threw away a piece of brain that was supposed to be dissected and used in the days to follow. A few hours before, the Italian scientist Camillo Golgi had thrown some silver nitrate in the same trash can. The morning after, Golgi salvaged the piece of brain from the trash can and observed that the nervous tissue had absorbed the colorant – neurons had become visible by turning black. Almost by accident, Golgi discovered a way to identify neurons by applying coloring substances to nervous tissues. After this discovery, the Spanish scientist Santiago Ramón y Cajal affirmed that each neuron is a standalone anatomical unit and that there is a space between two neurons. The two scientists shared a Nobel prize in 1906 for the discovery of neurons.

Dendrites, small extensions of the cell, and the axon, a bigger extension, branch out from the neuron cell. The former receives incoming signals, and the latter process the outgoing message. These connections, called synapses, were discovered by the English physiologist Charles Scott Sherrington. Synapses are not a physical connection between neurons as there is always a microscopic space between them. Therefore, the signal has to change from electrical to chemical to cross this space. The axon also releases substances, called neurotransmitters, which are gathered by receptors on the membrane of the receiving cell (output neuron).

Without delving too deeply into this topic, let us use the example of a suggestion prompter to understand the importance of neurotransmitters. Adrenaline, for example, is a fast neurotransmitter and it has the role of prompting an immediate response – same as a feline's sudden movement in front of danger. Somatostatin[51], a slow-acting neurotransmitter, is able to induce long term changes by inhibiting specific neurons. In the next chapters, we will use mathematics to explain these easy concepts.

In short, the basic elements of a typical neuron are (Figure 2.3):

- Soma. This is the cell body, which is roughly spherical (dimension: 5–100 μm).
- Axon. This is an extension of the soma. The axon has a length varying from a fraction of a millimeter to a meter in the human body.

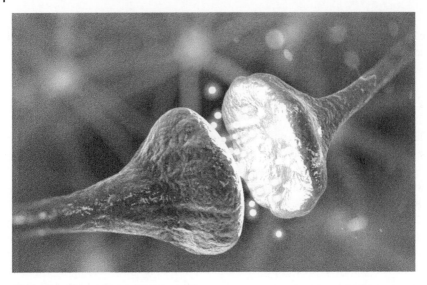

Figure 2.3   The mystery of a nerve cell.

- Terminal buttons. At the other end, the axon is separated into several branches.
- Dendrites. Another kind of extension around the cell body, they branch out in the shape of a bushy tree.
- Synapses. Terminal buttons are placed in special structures called synapses, which are junctions between terminal buttons and dendrites. A neuron typically drives 103–104 synaptic junctions.

The biochemistry of the human brain is rather complex and, if you are interested, there are recommendations to other works on the topic in the source sections.

It is true that there is a lot about the inner workings of the human brain that we are yet to understand but, to a basic level, we can say that human neurons get a number of inputs and outputs on a sensory level.

2.2   From Biology to Mathematics

In the Table 2.1 you will find comparisons between a biological network and an artificial network. It is important to summarize the main differences so as to avoid confusion from now on.

Table 2.1 Comparison between main characteristics of Biological and Artificial Intelligence.

	Biological	Artificial (as of 2021)
Size	The human brain contains about 86 billion neurons and more than 100 trillion synapses/connections.	The maximum number of "neurons" in artificial networks is usually in the ballpark of 1000.
Topology	Neurons can fire asynchronously, with a small portion of highly connected neurons (hubs) or a large amount of less connected ones.	All neurons are connected to the ones closer to them and can pass signals synchronously.
Speed	Certain biological neurons can fire around 200 times a second on average.	The speed of an artificial neuron depends on the speed of a computer processing up to 3 GHz (3000 times per second).
Fault tolerance	Information is stored redundantly so that minor failures do not result in memory loss.	Artificial neural networks are not modeled for fault tolerance or self-regeneration. Artificial neurons and their connections stay intact.
Power con- sumption	An adult brain operates on about 20 watts (barely enough to light a bulb).	A single Nvidia GeForce Titan X runs on 250 watts.
Learning	Brain fibers grow and reach out to connect to other neurons, and this neuroplasticity allows new connections to be created. Synapses may strengthen or weaken based on their importance.	Artificial neural networks have a predefined model, where no further neurons or connections can be added or removed.

2.3 We Came a Full Circle

So far we have been inspired by biology in discussing neural networks and how these are similar to our brain's function. Now we shall see how neural networks can help us solve problems of biology. Recent research on SARS-CoV-2 (Covid-19) has brought up an old issue in molecular biology: protein folding, which happens when proteins obtain their three-dimensional structure. This structure can take on 10^{300} different configurations starting from the same sequence of amino acids (molecules that make up proteins). To predict a protein's configuration means going one step toward understanding the way they work. This configuration is the result of complex biological interactions, which are also difficult to recreate. Traditional procedures, such as *three-dimensional crystallography*[19], require a year, if not longer, to be carried out. Therefore, one can only imagine the amount of time needed to predict the configuration of all known proteins.

Advanced computing technologies, such as distributed computing, are used at research laboratories worldwide to solve this issue. Some of these make use of recursive[45] and *attention-based convolutional neural network*[8]. Without delving too deep into this topic, let us just acknowledge how neural network training on a series of protein configurations has brought us to the prediction of the key protein of Covid-19. This valuable example is in the work done by DeepMind with their project AlphaFold[1]. Other projects by DeepMind are worth a mention, such as AlphaGo[1].

2.4 The Model of McCulloch-Pitts

The very first model of an artificial neuron was the Threshold Logic Unit (TLU) proposed in 1943 by Warren S. McCulloch (1898–1969, American neurophysiologist) and Walter H. Pitts Jr. (1923–1969, American mathematician).

Although quite simple, their model has proven extremely versatile and easy to modify. Variations to their original model have today become the basic building blocks of most neural networks.

In essence, given a series of inputs x, their linear combination (sum) produces two pieces of information that are different and will represent two opposite concepts. This binary approach[6], also considering the technology available at the time, can produce a true/false kind of result when we are trying to identify objects.

$$\sigma(x) = \begin{cases} 1 \rightarrow if \sum_{k=1}^{n} x_k > \Theta \rightarrow the\_image\_is\_a\_cat \\ 0 \rightarrow otherwise \rightarrow the\_image\_is\_a\_dog \end{cases}$$

The limit value Θ in the expression is not casual but defines very clearly the set that a specific data belongs to. If we wanted to tell a dog from a cat by using this model, we would need to put the two animals in different categories. Each category is defined by the value Θ. Therefore, once we have processed the distinctive elements of each animal (weight, size, speed, diet, and so on), the calculation will give us one answer: either dog or cat. It is clear that the limitations of this model are visible when we try to discern objects in a more detailed way and the values 0 and 1 are not sufficient to do so anymore.

2.5   The Artificial Neuron of Rosenblatt

Let us say we want to create an algorithm able to classify some photos to determine if the image depicted is of a cat or a dog. Our brain is able to determine the difference in an instant, but how does that happen? Our brain combines a series of inputs, such as visual, acoustic, and olfactory and compares them with what is stored in our memory – for example the image of a cat or a dog (ideas).

While we are not qualified in this instance to explain the functioning of the human brain, we can explain how an artificial network would see things (Figure 2.4).

Almost fifteen years after McCulloch & Pitts, the American psychologist Frank Rosenblatt (1928–1971) invented the *perceptron*, a major improvement to the previously proposed neuron model. This model (Figure 2.5) showed that artificial neurons can actually learn and store specific information.

A *perceptron* is composed by the following parts (discussed in Part III):

- Input values (X), weights (W).
- Sum, activation function.

The modeling proposed through the use of the *perceptron* proved to be the solution to an initial classification issue, while it could also have been solved with statistical analysis.

From the Figure 2.6, the model schematization shows some similarities with the biological neuron shown in the first part of the book. The cell body, the connection (axon and dendrites) with other neurons and the terminal (axon) are fundamental features for transferring and processing all the information needed to estimate a pattern[44] or a certain logical sequence within a set of initially independent data.

Figure 2.4   How is a cat seen by a neural network? Using neural networks to classify images. Source: Shutterstok.

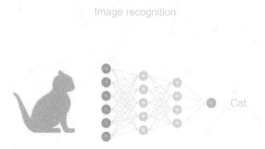

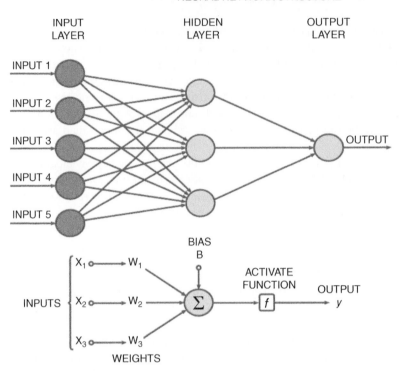

Figure 2.5 Components of a perceptron: inputs and weights. Source: Shutterstok.

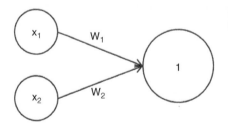

Figure 2.6 A perceptron: two inputs and two weights.

The possibility of simultaneously combining all incoming values with weights (W) allows us to use the information toward the final solution. A common approach is therefore more effective than a statistical method.

Nevertheless, we have observed how the *perceptron* method proved to be disadvantageous.

In the Figure 2.7, we approach the issue from a mathematical and geometric point of view. This approach might clarify what is meant by saying that the network can classify or make predictions starting from a batch of data.

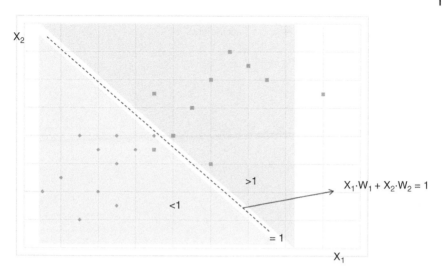

Figure 2.7 What is a hyperplane? Here it is in a two-dimension representation.

A *perceptron* can identify a hyperplane[23], therefore more *perceptrons* can identify more than one hyperplane. Each hyperplane divides the data in one or more
hyperspaces[24].

It might seem easy to classify the information in two different groups or separable space[25], with values that are respectively over or below 1, over or below 0,
and so on. Nevertheless, it is not always straightforward to define two different
regions, especially when data is not presented in a neat distribution. Therefore, it
is not granted that we can apply linear split to the input information.

Let us go back to the images of the leopard and the cheetah in Figure 1.3 of
Chapter 1 and use the example of image classification.

Pixels[31] are defined by numbers. Each number represents a color (e.g. RGB
numbers[40]). These numbers can be used as the input layers of a perceptron
(Figure 2.8).

The solution to our problem is to put together more *perceptrons* to define a more
accurate hyperspace region (Figure 2.9).

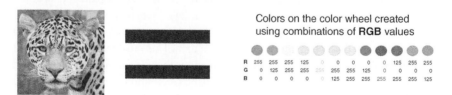

Figure 2.8 Simplified computer vision approach.

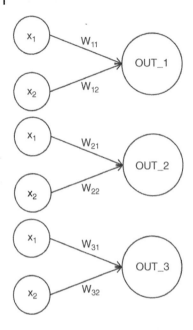

Figure 2.9   Three perceptrons: two inputs and six weights.

If we take more than one hyperplane into account (Figure 2.10), we can better define the hyperspace. Nevertheless, we will have not classified the depicted animal, as each *perceptron*, and therefore each hyperplane, defines a limit but not an exact value. A *multilayer perceptron* (MLP) (Figure 2.11) generates a set of non-linear functions with unknown parameters (weights). This set of functions, as we will see, are designed to better train the network or as suggested in other sources, to better fit the input data.

The output of the elaboration will be a binary value 0 or 1, depending on whether the original image is of a leopard or a cheetah. This is what we call binary classification. A network can actually classify different types of outputs but let us start with easier concepts for now.

If we look at the data distribution, we can observe that some colors are very similar to other colors (there are many nuances of red and blue for example). On top of this, it can be exceedingly difficult to place them in a specific position inside the image if the animals being classified are already very similar (leopard or cheetah?).

The magic happens in the so-called hidden layer, where weights are continuously updated as the network is trained. Training a network means to build a model that can correctly generalize the input data. This must tend toward the real data that correctly defines an animal. It is obvious that we cannot aim for absolute accuracy, therefore we have to use probabilistic calculations, as we will illustrate in Chapter 5. As of now, we can assign a specific function (continuous or derived) to each neuron - this way we can allow the network to develop a model as close as

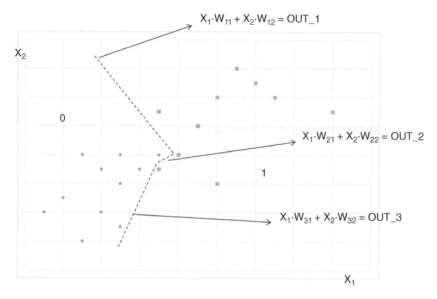

Figure 2.10 Are three hyperplanes enough to define one or more hyperspaces? Two hyperspaces (0 and 1) defined by three hyperplanes (OUT_1, OUT_2, and OUT_3).

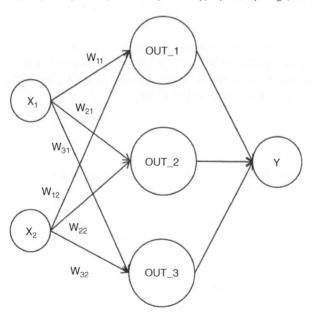

Figure 2.11 MLP – Multilayer Perceptron. Conventional sketch.

possible to reality. Through this function, we will get closer to the desired output, that is to say 0 or 1.

A neural network with a sufficient number of hidden neurons can adapt to any non-linear functions. Most physical phenomena are not linear, and we cannot convey them through linear combinations of their variables. For example, from a mathematical point of view, the model of a physical phenomenon could contain algebraic equations with at least one variable different from another (x^2, x^3, and so on). This aspect makes the search for mathematical models a difficult one. In our applications, we generally simplify a non-linear problem as a linear problem, resulting in a linear model. We avoid this step when using neural networks as we can take advantage of being able to perform the aforementioned simplification.

The main aim of a neural network model is to find the best compromise between accuracy (*fit*) of the training data and the ability of the network to generalize or function correctly on a new set of data.

Now we can define the following set of data.

- **Training set**: a set of values from pictures of leopards and cheetahs (as RGB values of the pixels), associated with their binary value (0 for leopard, 1 for cheetah).
- **Test or validation set**[54]: a set of values from pictures of leopards and cheetahs to test the accuracy of the network, once the model is trained.

The training process involves finding a set of weights in the network that proves to be efficient at solving the specific problem. This training process is iterative, meaning that it progresses step by step with small updates to the weights. Each iteration defines a change in performance of the model.

2.6 Final Remarks

In this chapter, we have introduced some descriptive criteria for our neural networks. These rules will be useful for understanding some mathematical operations that will be explained in each chapter until we reach the network learning phase. Let us go to the fundamentals of neuroscience. It has been affirmed, maybe erroneously, that memory resides in the alteration of synapses happening between neurons.

It is not easy to describe how, for example, the distance, duration and/or frequency of these connections are measured. This is because most of their properties are found in molecular processes that are not easy to interpret.

From now on we will take a risk and assume that our brain is a computer made of neuron bits. We will build connections between them and define their properties one by one.

Let us imagine that N input neurons are available to us. Connections between neurons are characterized by a (scalar) number that we will call a weight (W). Weights are combined with the value associated with each input neuron, as long as a connection is produced. In case this connection (symbolized by an arrow) does not occur, then no mathematical combination will be produced.

Let us start by writing the (linear) combination for a single connection (Figure 2.12), where for each weight we multiply the value of the neuron. We label the *n-th* neuron with *n*.

$$W_{11} \cdot x_1$$
$$W_{12} \cdot x_2$$
$$\vdots$$
$$W_{1n} \cdot x_n$$

Figure 2.12 Neurons connection – Conventional sketch.

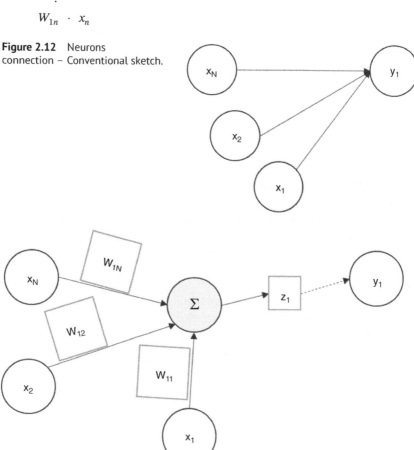

Figure 2.13 Weights represented in a conventional sketch of Neural Network.

The (linear) combination for all neurons x and weights (W) is the following (we sum up each connection's combination). We label the total number of input neurons x with N. For ease, we will call this combination z.

$$W_{11} \cdot x_1 + W_{12} \cdot x_2 + \dots W_{1N} \cdot x_N = \sum_{n=1}^{N} (W_{1n} \cdot x_n) = z_1$$

In the following Figure 2.13, there is an outline of this first step. The link between z and y has willingly been indicated with a dotted arrow, as there is a missing piece in that area of the model that we will define in the next chapters.

2.7 Chapter Summary

Neural networks are a subcategory of AI, the same as machine learning, deep learning and computer vision. A network is a group of interconnected elements constantly exchanging information, exactly as the human nervous system which is made of special cells connected by interacting nervous fibers. The neural network is usually inspired by some aspects of biological neural networks. The human brain is made up of 86 billion nerve cells, called neurons.

Artificial neural networks have a predefined model, where no further neurons or connections can be added or removed. They are not modeled for fault tolerance or self-regeneration in the same way that our brains are. Frank Rosenblatt (1928–1971) invented the perceptron, a major improvement to the previously proposed neuron model.

A MLP, the most widely used neural network for classification and regression problems, generates a set of non-linear functions with unknown parameters (weights). These functions are designed to better train the network or as suggested in other sources, to better fit the input data. The magic happens in the so-called hidden layer, where weights are continuously updated as the network is trained. Most physical phenomena are not linear, and we cannot convey them through linear combinations of their variables.

The main aim of a neural network model is to find the best compromise between accuracy (fit) of the training data and ability of the network to function correctly on a new set of data.

Questions

1   How do we reach the correct integration of AI?

2   What are the inputs and outputs of a neural network? List what SE processes can work out in the input and output activities of a Neural Network.

3 Why can a Neural Network be considered essential to avoid redesigning the existing product/project?

4 How is the data analysis, associated with the first phase of a Neural Network, comparable to the choice of requirements for any SE technical process?

Sources

Brady, S.R., Sigel, G.J., et al. (2011). *Basic Neurochemistry: Principles of Molecular, Cellular, and Medical Neurobiology*, 8e. Elsevier.

Loiseau, J.-C.B. (2019). Rosenblatt's perceptron, the first modern neural network. Medium, Towards data science (online resource). http://www.ai-shed.com (accessed 07/01/2021).

Nagyfi, R. (2018) The differences between Artificial and Biological Neural Networks. Medium, Towards data science (online resource). http://www.ai-shed .com (accessed 10/01/2021).

Jumper, J., Tunyasuvunakool, K., Kohli, P., Hassabis, D. and AlphaFold Team. (2021). Computational predictions of protein structures associated with COVID-19, Version 3, Medium, Towards data science (online resource). http:// www.ai-shed.com (accessed 07/01/2021).

3

Engineering Neural Networks

In practice, the aim of artificial intelligence (AI) is to make machines capable of performing tasks, which require intelligence as handled by a human. AI based prediction models have a significant potential for solving complex environmental applications that include large amounts of independent parameters and nonlinear relationships. Because of their predictive capabilities and nonlinear characteristics, several AI based modelling techniques, such as artificial neural networks, fuzzy logic and adaptive neuro fuzzy inference systems have recently been conducted in the modelling of various real-life processes [...]

Frédéric Magoulès and Hai Xiang Zhao, Data Mining and Machine
Learning in Building Energy Analysis

In the previous chapters, we have mentioned numerous times the possibility of applying computing methods to our everyday life, be it professional or private. We have also seen how a systematic and systemic approach enables the interpretation of reality and such an approach can result in reliable evaluations. We have specifically investigated the use of neural networks to leverage in the decision-making process and defined the main characteristics of a computational algorithm. We now need to put these applications into context – integrating them in a system and trying to understand how a mathematical approach can regulate the system itself. To support this task, we will use a discipline that was born in the last century and is called systems engineering – we will rely on its capability to help us manage the complexity of a system. We define a complex system as a series of elements working together toward a common objective, or a variety of interacting parties carrying out the same task collectively. The word "complex" clarifies how the elements of the system that are not coherent with one another to begin with become essential to reach the scope through logical and interactive links. Systems engineering does not only include traditional engineering but also other branches of knowledge that operate in areas such as: management, social,

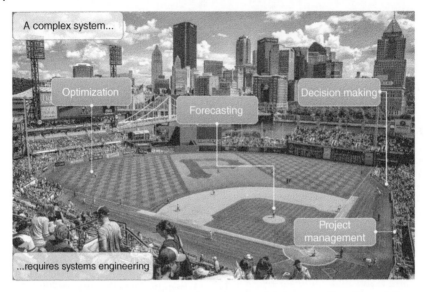

A complex system...

Optimization

Forecasting

Decision making

Project management

...requires systems engineering

Figure 3.1 Sports business as a field of application of Systems Engineering and Neural Networks. Source: Pixabay.

political, and legal fields – these all aim to help professionals in dealing with complex reality (See Figure 3.1).

Going back to one of the examples we have seen in the first pages of this book, the question that should immediately come to the forefront of our mind is: how can we develop an algorithm to tell a leopard and a cheetah apart?

Again, developing a new system is the result of more disciplines coming together (multidisciplinary approach), and the elements that are part of the system are themselves linked to other subsystems to guarantee the exchange of information. In the case of animals, we need to find a link between zoology and IT, statistics and veterinary medicine. Great technical variety together with a reasonable foundation of management skills and experience make systems engineering a valuable approach to problems of varied nature – with the aim of developing a cognitive process that is as reliable as possible. We believe that carefully following a detailed plan helps us reach our goal.

The human brain, and its cognitive side, can be influenced by mental pathways that might help us manage our judgments in an effective way or, on the downside, hinder our choices. Mental schemas, in cognitive psychology, define the way our thoughts are organized, and the way they behave in understanding the information and inter-relationships. Our goal is to create as many schemas as possible, so that these can help us with our actions, but above all we want to improve the way we make connections between different disciplines. If we are to bake a cake with the aim of serving our guests a tasty dessert, not only do we need the recipe and the quantities of necessary ingredients to bake the cake – we also need to know which ingredients we are going to use (raw materials). On top of that, we need to establish a production line and understand, for example, why the oven temperature needs to be at a specific level – these details will all help us reach our goal.

3.1 A Brief Recap on Systems Engineering

Systems engineering, as a process, was born at the beginning of 1900 in the communications and aircraft sectors. It developed quickly during the Second World War, as it contributed to the creation of systems that were more and more innovative and complex. At the beginning of the 50s, systems engineering was recognized as a profession and certified in the industrial field. The development of communications networks, ballistic missiles, radars, computers, and satellites, all labeled as systems, allowed this subject to be recognized as necessary and impactful. The systemic approach became commonplace in many fields, from social to technical. Standardized rules and regulations have been updated to integrate this subject clearly. Systems engineering has applied its features in order to solve complex problems. Thousands of engineers have declared systems engineering as their calling. Professional associations have created specific sections and published magazines based on the topic of systems and their development. Universities have established departments of systems engineering or programs oriented toward the development of this subject. Various books have been written on many aspects of the subject (Machol 1965, Chestnut 1967, Blanchard et al. 1981), until the moment when the profession was formally recognized by the founding of the International Council of Systems Engineering (INCOSE) (1990 records, National Council on Systems Engineering 1994, Ascent Logic 1995). For more information on systems engineering, we recommend the INCOSE handbook, also mentioned in the sources of this chapter.

3.2 The Keystone: SE4AI and AI4SE

In this book only the AI technique of artificial neural networks (ANN) is covered but it is enough to show the permeating use of this technology in most aspects of our lives. It is sufficient to pay attention to the emergent Intelligent Systems like autonomous vehicles in all their facets (cars, trains, submarines, aircraft, ships, etc.). It is also clear that both AI and Systems Engineering are powerful tools at our disposal to create simple and complex systems from our children's tree home to constellations of spacecraft for worldwide internet broadcasting.

What may be clear to you is that systems thinking can be used to create better, cheaper more capable products; what may be less obvious is how system engineering methodologies make more and more use of Intelligent agents to manage the overwhelmingly complex effort necessary to manage the systems engineering effort itself. All over the world there are teams of experts devoted to using the most advanced data analytics techniques to achieve that goal.

Therefore Intelligent systems are improved through system thinking while AI techniques are nowadays used to perform efficient systems engineering.

3.3 Engineering Complexity

We have established more than once that ANN are able to learn from experience, make assumptions from examples and identify the information in a large amount of data. These characteristics have increased the importance of ANN in systems engineering, especially in the last 15–20 years.

The system interface is the logical part that allows two or more systems to communicate and interact with each other. The connection allows data and information to be exchanged when the system is in equilibrium, i.e. when an interchange of information is generated between the individual elements that make it up, so that the system can interface with the external environment. If we consider ANN as a system in equilibrium, then we will be able to integrate it with other systems through its interfaces. Even if it is used superficially, the approach of systems engineering will highlight the concepts of "system elements" and "system interfaces" – overall, it will highlight the complexity of any kind of system built to examine and solve a given problem. Complexity is an intrinsic characteristic of a system, but also a regulating factor of its life cycle, in which the differences and analogies of all parts of the system are established. To understand and manage the complexity of a system prepares us for all those events that are difficult to foresee, such as cost overruns or planning delays.

We will define two fundamental aspects of a system: dimension and connectivity (Figure 3.2). These two aspects define the "structural complexity" of a system. Dimension is linked to the number of elements in the system, while connectivity is about the type of connections. There are other kinds of complexity, but structural complexity allows us to introduce the most straightforward application of neural networks.

Let us go back to considering "system thinking" as a far-reaching multidisciplinary approach. Many applications in engineering not only deal with the design or production of a system, but also with the many possibilities to evaluate the system's **reliability**[2].

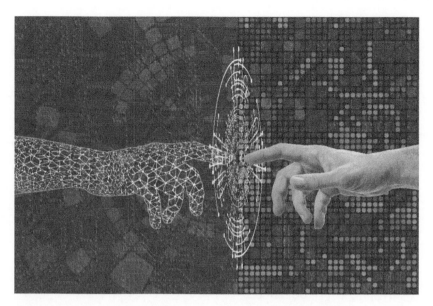

Figure 3.2 Where systems engineering and AI meet. Source: Pixabay.

As an example, let us use a generic autopilot system designed to control an aircraft autonomously. The data are processed during the flight, using a specific software which can create a specific route while managing the flight commands.

The system can be made of the following parts:

- GPS/IRU: *Global Positioning System/Inertial Reference Unit* provides a direct location of the aircraft. This data can be compared with the location of the area considered to acquire precise directions.
- BUS BAR: power supply system. All the elements are linked to the power supply system that is connected to a battery or to the aircraft's power generator.
- FMS: *Flight Management System*, which defines the flight route. This data is compared and processed together with that provided by GPS/IRU.
- CPU: central processing unit gathering all the data provided by the other components.
- SOFTWARE: data processor.

We have defined the main components of an autopilot system. Some of these components can be deemed as subsystems. The GPS/IRU (See Figure 3.3), for example, can be examined separately once integrated with the system under consideration. To derive the complete system architecture, we will need to know the input, such as the signals coming from the aircraft's fuselage, and the output, such as the position of the aircraft. Each subsystem can be dealt with separately and

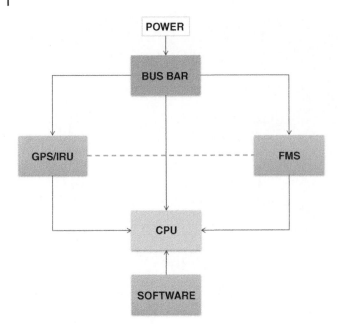

Figure 3.3 Example of System Architecture for an autopilot system.

in the same way we would consider the main system. The characteristics of a "subsystem" must be compatible with the purpose and the objectives of the main system; therefore, they have to be optimized by making use of specific methods. Now, let's try to think how an automated subsystem, integrated in a complex system, can become an autonomous subsystem. As autonomous systems we mean an "intelligent" system that can meet the requirements of a complex system (in this case an aircraft), is able to explore the environment and analyze events without human intervention. So, a loophole is sought that allows the systems engineer to be able to manage a new product that actually uses complex formulas and mathematical models to make quick decisions based on established parameters.

This example shows how the structure of a system starts from defining its objective, as well as the requirements and features of its components.

The approach employed by systems engineering not only helps us in the development and production phase of the aeronautical components, but also in defining a process to assess the performance of the product. The system engineer must not think of abandoning the best-known techniques for analyzing a complex system in order to integrate a new subsystem into it that allows to become it autonomous.

For example, a reliability allocation can determine the levels of detail required in developing a complex system. Therefore, an accurate analysis of system failures (FMECA-Failure Mode Effect and Critical Analysis) and Fault Tree Analysis (FTA) can improve its performance (RAMS-Reliability, Availability, Maintainability and Safety) and influence decision-making in case of unexpected events. A reliability analysis is definitely an ongoing iterative activity over the life-cycle system. This type of activity is required to update reliability estimates that can affect design, verification, and validation process.

The analysis process must be continuous and require in-depth experience on the complex system. It must never be forgotten that making a decision is no easy matter.

Imagine being on an airport land side, carrying out extra maintenance work on an aircraft which showed signs of failure in its autopilot system (aircraft subsystem) as soon as it landed. You have also been informed that the aircraft is scheduled for an overseas journey in the next few hours. You immediately start looking for the failure in the system and notice that the inertial system needs to be replaced (a subsystem element) as it is damaged. The solution to this problem is to replace the damaged part with a new one, but what if there are no new parts in the warehouse? The analysis of possible solutions will lead you to face the issue in the short time frame you have available, and you may decide to cancel the next flight. On the other end, a systemic approach, considering all the requirements to solve this circumstance, such as transfer times from warehouse to warehouse or the reliability of the aircraft components, would lead you to a better decision. An optimized decision-making process should assess the reliability of the primary components of the aircraft and optimize the warehouse management to have the necessary parts available in case of issues.

It is clear by now that making a decision is not as simple as it may seem due to the many variables at play and the amount of knowledge of relevant and non-relevant disciplines that is required. What is the link between the procurement, transportation and the performance (smooth functioning) of an aircraft (system)? This is one of the many questions you will need to ask yourself in the field of systems engineering.

How does decision-making work? For human beings, usually, this is a complex process and the reason for this is beyond the aim of this book. We can nonetheless say that the human brain, though guided by much sensory information, cannot acquire data that is not gathered by these senses. A computational process or a predictive analysis can help our ability to make a decision if they are well-integrated to our evaluations and reasoning processes.

A computational decision-making process is a far-reaching subject that uses math to evaluate a big amount of data and determine the most convenient solution or line of action.

We would need to write a completely different book if we wanted to delve deeper into this topic. Some interesting applications of the computational process in the financial field are asset allocation, security analysis, portfolio optimization, algorithmic trading, forecasting[33] and policy making. Some of these examples will be explained in the blog linked to this book.

The following paragraphs feature two original examples of decision-making, neural networks, and systems engineering. These are applied to sport business and the evaluation of corrosion on metal structures.

Business, as a field of work, usually addresses other specializations, such as accounting or financial analysis, while in sports it has a wider function, involving aspects like marketing and branding.

Corrosion instead is an extremely popular topic in aeronautics, as it is not always a foreseeable issue. Therefore, we think that the application of neural networks and systems engineering can make the forecasting process more reliable and take us from theory to practice.

3.4 The Sport System

Over the last 30 years, sports clubs have become real business ventures creating entertaining products and developing their own brand through a continued growth process.

In this section, we will present how systems engineering can, by means of various methodologies, analyze, program, simulate and help the decision-making process of a sports club. The aim is to optimize the main efficiency parameters, all the while supporting the management of the project. We can reach this aim by using new techniques to improve resource management. A sports club can be included in the category of complex systems since each of its sub-systems or, in general, each of its elements contributes to defining its mean characteristics: complex and multidisciplinary systems.

The ideas of complexity and system cannot be discussed separately, as they are interconnected. We cannot understand how a system functions without considering its complexity. Upon categorizing the sub-system disciplines, the system engineering approach allows to analyze each sub-system independently of the other ones. This approach is finalized to look for one or more shared points within the various sub-systems and then to find a sort of gateway for the information exchange. An optimized view allows the obtaining of a balanced system. So, we can correctly evaluate the complexity and the architecture of a system by discerning its elements and the way they are interconnected.

We will need qualitative techniques to determine the architecture of a complex system, such as a sports club. Analytical techniques alone are not sufficient to define the different levels of a system. As we progress to higher levels, we will need to use heuristic methods based on experience, abstraction, and integrated modeling.

By using these techniques, we focus on the essential elements of an issue to solve it easily. To model a system at the highest possible level, in the first instance, we exclude from consideration the minor system elements (detail level). We can define a level as a set of elements that are grouped through logic and experience. A high level is a set of macro units having the same benefit for the given system. The system can then be divided into lower and more simple levels, which can be dealt with one by one by only considering its input and output connections. A recurring phenomenon in the organization of multidisciplinary structures is a high degree of low-level interconnection, which can increase the complexity of the system (See Figure 3.4).

A sports club is a complex system that includes different subsystems cooperating toward a main objective and aiming to solve inter-disciplinary issues, such as the management of human resources and the interaction with the sports community – it is made of elements (stakeholders) that directly or indirectly affect or anyway belong to the same system. A sports club specialist is responsible for optimizing the characteristics of a particular system component. Specialists will try to optimize those aspects that they hold to be important. Reasoning by analogies, it means that the medical staff can manage the sports club as a hospital. The specialist can understand that the whole system is a group of balanced components but inevitably, his attention is necessarily focusing on those issues that relate most directly to his area of expertise and the responsibilities assigned during the life cycle of the system. On the contrary, the systems engineering must always focus on the system as a whole. While addressing issues of medical specialization the systems engineering must understand how they may affect athlete performance

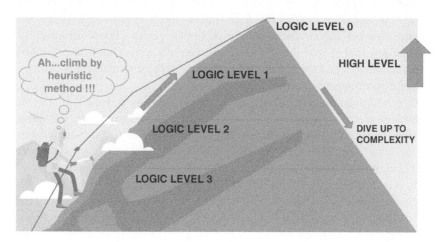

Figure 3.4 A system seen as a mountain – we have to climb this mountain moving through all logic levels.

and generally how it may affect the sports club objective. Other issues, such as risk, cost, or long-term feasibility of the system, can be treated since the complete analysis allows making a best decision especially in the convoluted cases where applying non-linear analytical methods is suggested.

Therefore, the best decision must take into account the balancing and consequently must come from a deep understanding of how the system works. It is this judgment that systems engineers must exercise every day, and they must be able to think at a level that encompasses all the features of the system. The systems engineer's perspective requires a different combination of skills and knowledge areas than that of a specialist or manager.

In addition, guiding the development of a complex system and achieving the intended objective necessarily implies that each of the components of the system receive the proper attention and resources. This often involves defining some technical trade-offs in order to reach a workable interface between the key elements of the system. A general view of a complex system ensures that the balance is such that each of its components has the opportunity to grow but not at the expense of another. For example, the performance of athletes can be improved even without large investments. As we have mentioned previously all the elements can seem independent between them, but it is necessary to find a correct balancing since a decision has to be made in accordance with the purpose of the system. Now we can provide an example application of systems engineering to a sport club, as a starting point to a larger discussion on the topic. In this first phase, we start from the main aim of the sports club – we combine the relevant subsystems, and we link them based on their technical specifications. It is also important to contextualize the system in its operating environment as its external interactions can influence the way it functions and can divert it from the set objective. The application of systems engineering to sports is a new approach, but a remarkably interesting one – it has the potential of reaching the main objectives of a sports club such as winning a championship, establishing a certain position in the international rankings or organizing business events.

Figure 3.5 depicts a sports club as a system of systems framework. Although the organization of a sports club may seem easy to understand superficially, it is indeed complex. Therefore, we need to build a system architecture that can define many interactions happening in its operational level.

A new and central role is introduced in the Figure 3.5: the project manager (PM). Why is the PM an essential position in sports clubs? What is his or her role?

First, let us define the word "project." What is a project in the sports business? When we decide to create a new product or improve an existing one, we need the input of many people with different skills to devote years of effort in the creation of a preliminary design. The size and complexity of these efforts require a team specifically dedicated to the coordination implementation. This is the

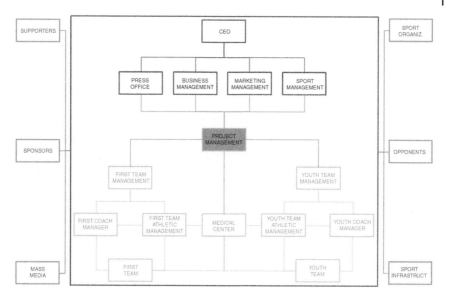

Figure 3.5 Architecture of a sports club as a framework.

definition of a project, which is led by a PM supported by a staff of professionals in the field.

Similarly, a PM takes care of a particular technical discipline, and he must be able to manage technical people. A systems engineer, on the other hand, requires significant capabilities in all three components, representing the balance needed to cover the needs of an effort for the system. In this way, the systems engineer operates in more dimensions than his or her peers do. Systems engineering can also help with the development and improvement or management of all those processes and methodologies used during the execution or integration phase of a system into its operating environment. Many processes, such as those belonging to the technical process, are dependent on each other and provide a complete view of the system lifecycle through explicit relationships shown between the requirements and needs established to validate the developed product. A sports club that we believe does not produce anything tangible such as an airplane or a missile has an objective. This objective can comply with a product through the proper evaluation and validation of management processes. At this point it is useful to think how a systems engineer can stay behind all these activities that among other things are proper to a complex system.

The engineering effort is optimized by establishing objectives, leading the implementation, evaluating the results, and planning the necessary actions to keep the project on track.

Systems engineering deals with three main areas: project objectives, deadlocks, and management of a project all round specificities. All these aspects are part of a set of constraints, financial or planning ones, that have to be developed and clarified by professionals from all the relevant sectors. Who puts this information together? You might have gathered that "interdisciplinary" leadership is necessary for complex systems to succeed.

The engineering of a new complex system usually starts with an exploratory phase during which a new concept of system is developed to satisfy a known need or to make use of new technology. In the sport business, project planning, financial and contractual performance, as well as customer relationships are very rarely implemented. Systems engineering is a holistic approach.

The PM has a key role in the implementation of any project. Their main goal is to harmonize costs, time frames, quality and, most importantly, client satisfaction.

Regardless of project aims, a good PM must be able to interpret the key objectives from the beginning to the end, making sure that the stakeholders' vision is realized according to their expectations.

A PM's activities can be divided into two main areas. One operational area, related to methodology, tools and techniques that work better for the project's aims. One relational area, related to interpersonal dynamics and communication. The role of the PM in the architecture of a sports club is imperative – they gather technical data (player performance), business information (profit, costs, investments), and project structure (processes, activities, life cycle, obligations, and requirements). Given the information, they make a decision to reach the final aim. To tick all the boxes, the PM needs a support tool based on a database (big data).

3.5 Engineering a Sports Club

Many sports clubs are led by a president or a chief executive officer whose figure has, since the 1900s, been the one and only stakeholder. From their influential position, they would financially support the team championships for it to compete against other clubs.

Let us look at a sports club from another point of view, namely as a complex system supported by more than one stakeholder, who can be part of the same system and be members of external organizations.

Recently, the management of a sports club has revealed that a vital factor for success is a good balance in the interests of all parties involved. In a sports club system, some stakeholders provide capital, and they expect a positive return on their investment. Supporters, employees, business partners and suppliers become stakeholders of the club, each of them having different objectives.

Figure 3.6 The basics of the
System Engineering approach.

A first step would be to link each requirement to a specific stakeholder: CEO, supporters, sponsors, sports organizations, mass media, managers, and players.

The sports club must guarantee an effective result (victories for example) based on the set requirements and use the result to reassure the requirements – this would give life to a controlled cyclical process (Figure 3.6).

It would be ideal to find a common denominator that satisfies all the stakeholders, directly or indirectly. We have assumed that victories would be useful to all stakeholders, but if we employ a heuristic process, we can find other paths to follow, as well as new facts. The heuristic process is one alternative way to establish some common goals.

With this in mind, let us establish that a bigger number of victories can not only satisfy the supporters and the team, but would also encourage the sponsors to sign remunerative contracts or mass media to put a spotlight on the club's achievement. Is there a link between mass media and supporters? The answer to this question might seem obvious, but would we be able to establish all the interrelations between supporters and mass media purely based on the club's objective?

3.6 Optimization

Let us find a part of the system that is essential to reach the set objective before delving deeper into the complex nature of the system. A system itself is composed of complex elements interacting with each other, which we can call subsystems.

Let us observe the subsystem linked to a sports action, where the performance data of a single athlete or an entire team is gathered and analyzed. Analyzing these details can surely help the club to get closer to its objective as an improved performance helps the club to win more matches during a football season, theoretically becoming more profitable.

One single athlete's performance is measured by a trained person (system component), during an exchange between the athlete and their manager. This will allow the establishment of the amount of work needed to obtain a specific level of performance in return.

In this architecture, we can define the "team manager," who gathers all the data provided by the given professionals and passes them on to the manager of the project. The latter must examine and establish, with the team manager, a strategy to resolve any issues that might have been raised during a match or a test session. Once the process has been defined, it is necessary to keep an eye on the costs and to communicate with the business professionals, as they will evaluate the investments required to resolve the issues.

Systems engineering requires a knowledge of the elements to be analyzed, with which one can develop a process aimed at performance improvement. To do this, we will define the main parameters required to create a mathematical model, which will help the decision-making process carried out by the sports manager.

Let's extrapolate the sub-system of interest (See Figure 3.7), we will have to ask ourselves about whether decision making should be maintained by an ANN. In order to do so, we can divide the performances in three main interrelated areas.

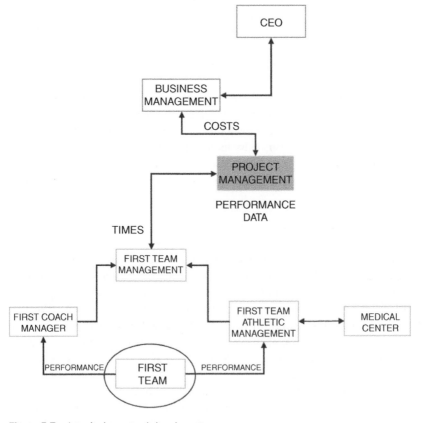

Figure 3.7 A typical sports club sub-system.

We can draw a spiral diagram or a scatter plot diagram – to simplify the process, we could also evaluate the dependencies in pairs. We will use a diagram similar to the fire triangle (oxygen-fuel-heat).

The following performances are only a small part of the parameters to be optimized, but they are a good starting point to create an appropriate intelligent algorithm for this example (Figures 3.8 and 3.9).

- Athletic performance: the physical ability of an athlete or a team in relation to the individual needs or the overall aims of the team.

  Evaluation process: the athletic performance level is evaluated during training or matches. The evaluation should be objective, based on the overall level of the team or individual level of each athlete.

Figure 3.8 Can this sketch also be applied to another business?

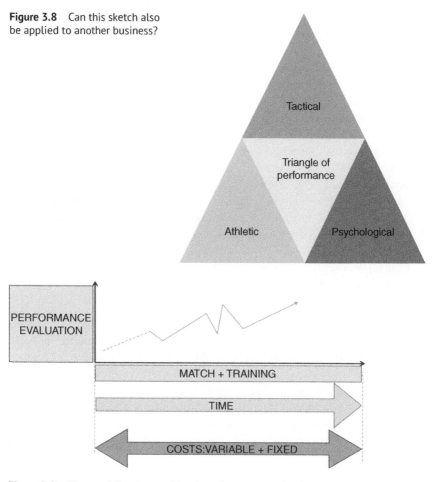

Figure 3.9 Time and Cost in an athlete's performance evaluation.

Time frame: match phase plus physical tests phase (life cycle).

Fixed costs: player and athletic trainer's salaries.

Variable costs: test costs, training, gear rental.

- Tactical performance: the strategic ability of the athlete, visible during game action, is verified through a series of behaviors (e.g. possession and movements), based on quick decisions. The tactical aspect is applied in sport situations defined by a high grade of uncertainty.

   Evaluation process: evaluating the ability of a player to apply the tactical rules given by their coach during matches or training by using objective data, based on the efficacy of movements and game strategy employed to reach the required result.

   Time frame: match phase plus tactical tests phase (life cycle).

   Fixed costs: player and coach's salaries.

   Variable costs: test costs, gear rental.

- Psychological performance (human factor): this factor is added to the process of physical training and the interpretation of new game strategies. Motivation is the main psychological ability of an athlete, as the basis of any successful sport by which the athlete satisfies its needs in terms of incentives, interest, and fun. The management of this parameter is allocated to a trained professional, while the athlete and the whole team are responsible for it too (self-critique). The need for self-affirmation and success (self-awareness) help to improve the performance. On the other hand, boredom and dullness lead to tiredness very quickly.

   Evaluation process: study of the athlete's personality, psychological basis of their motor abilities, competition preparation, athlete selection, group psychology, individual sports, training and competitions, and optimal performance mind-set.

   Time frame: match phase plus psychological tests phase (life cycle).

   Fixed costs: athlete and doctor's salary.

   Variable costs: test costs, gear rental.

In the next chapter, we will examine a specific case of performance evaluation, which will be useful to improve all the abilities of an athlete, by predicting trends or by choosing the best athlete based on data gathered by using a mathematical model.

With this example, we have established that, once the architecture of a system is defined, we can establish an evaluation process for a specific requirement. To comply with the requirements is essential to reach the final objective, this is only obtainable by optimizing the available resources, according to the performance triangle.

3.7 An Example of Decision Making

One of the aims of systems engineering is to manage the life cycle of a system (cradle to grave) effectively and consistently and, in general, develop the requirements of a system together with a strong management of the stakeholders. As we have seen in the previous paragraphs, a system (functional or physical), even if complex, can be managed through knowledge of its main elements. Through consistent supervision and proper management, the system will work correctly until the set objectives are reached. There are several management techniques for tailoring a system; in some cases, a **mathematical approach** may be used to produce useful result for decision making or for risk assessment.

In general, a decision-making process, applied to an in-service system, often takes into account the high degree of ambiguity, uncertainties, and trade-offs: it is precisely in these situations that we can appreciate the decision making approach.

Breaking down the issue into smaller, more manageable issues, allows us to focus on what is really important and then allows the development of consistent solutions.

Operational needs must be clearly understood in the logic of performance requirements, and therefore a decision database can be evaluated during the design phase. Operational analysis models, suitable for testing the validity of the in-service system, are provided by the link between operational requirements and basic requirements.

Leaving behind the sports business example, we have previously mentioned the autopilot, a system used to control the trajectory of an aircraft, marine craft or spacecraft without requiring constant control by an operator, and we have introduced some of the requirements needed to define its main components. The result should be a set of traceable requirements that can be used in the planning phase, in the procurement phase, and during reliability checks.

Over the last few years, there have been increasing engineering efforts aimed at regulating and managing the issue of metal corrosion. This phenomenon is predictable when, in the planning phase, a decision is made to use metal alloys. Corrosion is influenced by the environmental circumstances in which the structure (in-service system) operates, and protection techniques utilized (e.g. anodizing).

We think it is interesting to use this example to highlight the use of neural networks in decision-making processes resulting from a complex phenomenon such as corrosion.

Let us use a *V* diagram (Figure 3.10) to define the characteristics of the process by which we will manage corrosion as a phenomenon and over a specific timeline.

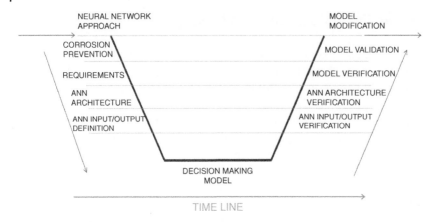

NEURAL NETWORK
APPROACH

MODEL
MODIFICATION

CORROSION
PREVENTION

MODEL VALIDATION

REQUIREMENTS

MODEL VERIFICATION

ANN
ARCHITECTURE

ANN ARCHITECTURE
VERIFICATION

ANN INPUT/OUTPUT
DEFINITION

ANN INPUT/OUTPUT
VERIFICATION

DECISION MAKING
MODEL

TIME LINE

Figure 3.10 V-diagram for NN in the maintenance support.

As with any process, such as the management of a possible breakdown (failure analysis), it is essential to understand the different interfaces involved in the system, such as a metal structure. A strict process control ensures that the system keeps on satisfying the initial requirements, which have been set for the entire life cycle. We need to know the characteristics associated with each part of the process, as they are essential to amend the previously established process each time new information is acquired. The configuration and change management of the system are also important steps, as they contribute to understanding the performance of the process. Change management deals with the desirable changes of the systems through its life cycle while configuration management deals with maintaining consistency of the system performance.

On the other hand, we use a solid conceptual architecture, established at the beginning, to create a process that is clear to all the stakeholders.

In this example, we want to create a process that monitors, in specific time frames, the onset of corrosion on metallic structures. The introduction of neural networks provides a valid technical support which can be managed with a V diagram if correctly integrated. Each part of the diagram has to confirm and substantiate the relevant characteristics of the process and it allows to obtain a correct management of changes. For more information on the V diagram please refer to the blog connected to the book.

What happens if our objective is to manage a system that is already designed, i.e. manage the process?

As we have mentioned at the beginning of this section, the approach used by systems engineering helps us understand the functional characteristics of a given system. When it comes to corrosion, it is clear that it would not be convenient to re-design the structure from scratch to avoid corrosion as this involves complex and costly analyses. Therefore, we can set our objective to prevent corrosion and put off the replacement of the damaged parts.

To summarize, this process will allow us to manage and implement a maintenance program based on the following rules:

- Defining the system's objectives.
- Defining the functional requirements and external constraints.
- Evaluating the elements of the system and its parts to set up and control possible changes.

The final objective is to reduce the costs of the project. Systems engineering helps the managers to provide a valid alternative to re-design, which could involve expensive investments on the constructor's side. In an operational phase, we can also reduce risks of breakdown and consequent reliability loss of the complex system. **Corrosion can be stopped if it is detected at the right time** (Figure 3.11). Imagine that we have established a maintenance task (e.g. inspection) for a point in time following the onset of corrosion, based on the estimates made in the design phase. We might find that corrosion is already quite extensive. Therefore, if we can foresee or prevent an irreversible situation, we can then guarantee a certain level of safety of the system.

Let us build the architecture of a corrosion management system, keeping in mind what we have said about the mathematical model of a neural network.

In the following picture, similar to the V diagram, we can see that the cyclical process can be stopped when we reach the desired level of reliability of the corrosion management system (Figure 3.12).

Finally, we can define a first process management based on the following:

- Defining the elements, clearly and correctly, of the whole system during its development.
- Defining the design characteristics and the system configuration.
- Acquiring more information during the operational phase of the metal structure.

OBJECTIVE	REQUIREMENTS	CONSTRAINTS	DECISION
Corrosion Control	Maintenance Task	Operational environment	New Maintenance Task
	Engineering Support	System Configuration	Re-Design
	Structure Design Specification	Operational loads	Repair
	Manufacturing Specification	Design loads	Preservation
	Maintenance capability		New Validation

Figure 3.11 What would be a set of generic requirements for a corrosion control system?

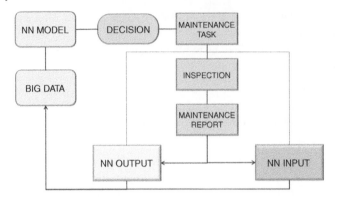

Figure 3.12 How can we use neural networks to improve inspection reliability?

3.8 Futurism and Foresight

As extensively argued, neural networks process a series of data and combine them through algorithms. On the other hand, in this chapter we have taken a different approach, slightly more abstract, by defining how the system and its elements (cause) determine the final and optimal result (effect).

According to the Institute for the Future, forecasting and foresight are disciplines that use various techniques to predict future events based on a large amount of data, by linking different types of data and identifying trends or unexpected connections.

On the other hand, futurism uses a less analytical and more conceptual approach, by recognizing the importance of qualitative prediction techniques, e.g. the Delphi method, and it draws on the experience and knowledge developed over the years.

As the following Figure 3.13 shows, foresight determines a long-term framework, but it also plays a role in quantitative (prediction) and qualitative (experience) contexts, similarly to what happens in war simulation games. We cannot overlook one or the other, but we have to take them both into account as they both offer their own input.

Forecasting is about making exact linear evaluations of future events that might have uncertain results to start with. For example, estimating the exact number of times flooding will happen in a specific area and in a specific time frame is a forecast. Even though this type of forecasting has never been the focus of neural network research, this has been consistently applied to prediction issues during their development.

Neural networks for forecasting operate on the basis of principles that differ from conventional methods and have more effective cognitive abilities – modeling

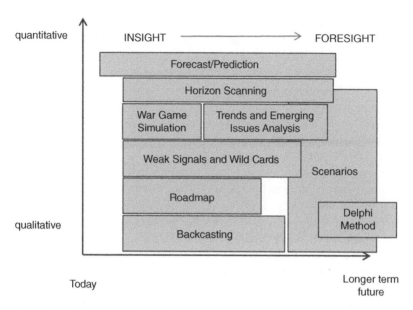

Figure 3.13 Futurism and Foresight. A summary chart. For more information on the techniques mentioned above, please visit: https://www.iftf.org/home.

of sensations and perceptions, models identification, models learning and memorization with the aim of identifying data knowledge. These models have a wide range of applications when compared to other synthesis methods such as fuzzy and conventional neural networks. Hence, we need to highlight that these models can be adapted through integration of a conventional neural network or fuzzy methods. Initially, we can set out a series of quantitative and qualitative parameters, then the model can be modified based on historical, cultural, psychological, and experience-based characteristics.

This book focuses on feedforward neural networks, but other types of neural networks are quite common when it comes to forecasting. For instance, convolutional neural networks[39] have become very popular for the aim of image recognition. Recurrent neural networks, especially gated recurrent units (GRU), are used for deep learning of time series. Also, spiking neural networks (SNN) can be well suited for robotics.

3.9 Qualitative to Quantitative

Let us go back to the sports club subsystem example concerning the athletes' performance. We wonder how to improve their performance or how to assess it.

Through newly developed tools, we can easily measure athletic and tactical aspects but not the psychological ones. So how can we combine them? How do we integrate the psycho-physical aspect in our analysis? We have mentioned in the first chapters of this book that a mathematical model might help us focus on a specific event and evaluate its characteristics. Would we be able to carry out this task on our own? Would we be able to predict the variations of all those aspects that are necessary to reach a set goal? We would know if we had a crystal ball.

The word Delphi was not casually chosen – it refers to the famous oracle used to predict one's future, with the condition that the questions would be well structured to interpret the answers. One key step in this methodology is the actual structure of the question, so that this can be used to meet the needs of our daily life. Which questions? Which answers?

With this methodology, we can evaluate the opinion of many experts concurrently and obtain operational directions that can be used as a reference by a single professional. In the last chapters, we have established how to mathematically define the bias and we have deduced that its function provides a certain degree of prejudice with regards to past evaluations. Prejudice can be a cognitive shortcut that can influence the expert's opinion or observation. The final verdict can be inaccurate. Differently from objective methods where the bias is introduced by the expert and their knowledge, the process used by the Delphi method depends also on an impartial judgment. Therefore, it is appropriate to let everyone evaluate the biased influence on the judgment expressed by third parties. Even though we have hundreds of different prejudices, we must select the ones that might have a negative impact on the quality of the analyzed results. In the industrial sector, a lot of attention is put on prejudices that can influence risk perception. We can use statistics to reduce the bias or apply methods of randomization. Randomization is a control method through which a researcher makes sure every subsystem of the population can be selected.

Let us go back to systems engineering: we can manage the complex and critical issues found in a system through the results of a systemic process and set rules. For example, we have to establish a series of improvement actions for a specific manufacturing process or for an athlete's performance. Through a questionnaire, we can ask a group of experts which actions we have to consider for improving performance. To answer the questionnaire, we will set a series of rules and instructions which will be useful to allow everyone to express their opinion. Once the answers are gathered and analyzed, we can identify convergent and divergent points of view. The aim is to gradually create a consensus among the answers we have received. Once the consensus is reached, or at least a balanced opinion, we can produce a common answer to be expressed statistically.

If we use multiple choice questions, we assume that the respondent holds a deep knowledge of the topic. We have to take into account a lack of knowledge on the respondent's side, which will appear in the subsystem of divergent answers.

Divergent answers can be discussed by all the experts to establish possible amendments to the questions.

We can divide the Delphi method into three phases:

- **Exploration phase**: based on the set objective, we choose the question we need an answer to. A well-structured question makes it so that the objectives are crystal clear, allowing us to have a defined idea of the issues we will be facing. Questionnaires are open ended and gather qualitative information. This is a vital phase as it establishes the overall quality of the project. We can use intuition and personal experience to reach a conclusion.
- **Analytical phase**: we analyze all the common aspects of each question that should have been highlighted in the previous phase. The questions are then grouped based on specific content to facilitate the following evaluation phase. This phase follows the processing and breakdown of the question we have previously described. It is based on the analysis of the levels of aggregation of each answer. In this phase, we can use self-awareness to reach the right conclusions, by employing inductive reasoning linked to experience.
- **Evaluation phase**: here we gather and analyze all the answers provided by our experts. They evaluate the topics and offer guidelines to their relevant group. In this phase, we must consider the role of cognitive bias. The experts must be able to review, evaluate, and critically develop their opinions.

This method can be ineffective if we use too many words to evaluate a simple event, even though a detailed description would improve the consensus level for complex events. It is therefore necessary to moderate the experts' opinions by defining key words in an initial phase. By applying this method, we can extend the questions to respondents who are less knowledgeable on the matter. It is useful to establish a discussion between experts and non-experts.

How can we use non-aggregated answers (outliers)? In the next steps, we will consider all uncertain situations in which aggregation levels might have not been reached. Between one step and the other, the results have to be shared with all respondents, whether experts or not, so that they can be aware of the other answers. This phase will allow the experts to critically review their position and evaluate whether they need to examine their answers again. This step will also avoid a disorganized discussion among experts, preventing time loss (as opposed to brainstorming).

In conclusion, the main advantages of the Delphi method are:

- The chance to have a high number of experts available to participate even though they might be geographically far from each other.
- A more effective evaluation due to time constraints to obtain answers, therefore avoiding the consequences of group dynamics, mutual influence, and leader pressure.
- Cost reduction.

We have to consider that there will also be pitfalls to this method. Empathy could result in lower attention to the answers or lead the respondents to influence each other. It is possible to misunderstand questions that are not too clear, resulting in a difficult summary process. The PM has an important role in this method too.

3.10 Fuzzy Thinking

How can we evaluate, in quantitative terms, those aspects that are only measurable on a quality level? Can you express your mood in numbers? We might have found ourselves in front of the following question: from 1 to 10, how much do you like this cake?

First off, logic is the ability of our brain to reflect and think, and it is a means to discern correct answers, but it cannot quite replace creativity, as we have previously noted. In other words, logic can help us organize words and sentences clearly, but it cannot determine which ones to use in a specific context. Logic to human beings is a way to develop a reasoning process in a quantifiable way, and a process that can be recreated and manipulated through a set of rules.

An evaluation is based on our perception and understanding of a particular physical phenomenon such as the taste of a cake. How exact can a personal evaluation be? Surely it could be altered by the inaccuracy of our reasoning, as we lack enough data or knowledge about baking. This inaccuracy is still a source of information, and it is needed to validate our reasoning: "this cake is good," "this cake is amazing." This kind of information is not precise, nor is it analytical. The ability to use this kind of reasoning for complex problems is the criteria used to evaluate the efficiency of what we call "fuzzy logic." This logic cannot solve problems that require a high degree of accuracy, but it can only express a first judgment on all those aspects that have a fuzzy meaning, at least to those evaluating them for the first time.

When talking about engineering models and products, the requirement for precision results in high demand of resources, such as costs and time. This happens during the product development phase and during the production phase. Equally, during the product or project management phase we will have expenses that are proportional to the accuracy of the validation process itself: the higher the accuracy, the higher the cost. In general, using fuzzy logic for a given problem forces a professional to confront the inaccuracy coming from results obtained with this method. We can therefore say that the fuzzy method is a general approximation tool for behaviors of a system in which there are no analytical functions or numerical relations that qualify it. Where necessary, we can use the fuzzy method to analyze algebraic systems, in fact one of its main advantages is to help the understanding of new or evolving problems.

Let us go back to our sport business case scenario, focusing on the psychological and physical state of an athlete. Through a quantitative defining process, we will convert the psychophysical level determined as high, average, or low (we can add more!) into a number. We will need to remember some notions of linear algebra to effectively lay out and explain fuzzy logic.

From now on, we will use A to refer to physical features and B to refer to psychological traits, while we will use C to refer to these two combined. Each set is represented by values (between 0 and 10 or between -10 and $+10$) established by our experts through a Delphi approach, with the aim of representing all features with numbers.

Input:

$$X_A = (x_1, \ldots, x_n)$$
$$Y_A = (y_1, \ldots, y_n)$$

Output:

$$Z_C = (z_1, \ldots, z_n)$$

The (x, y) values will become the input data of our model, while z will be the output data.

As for the psychophysical aspects, let us match a word to each athlete which will describe them: adequate, sufficient, inadequate (we can add more!).

The mathematical principles associated with fuzzy logic represent knowledge on a degree of membership, i.e. they convert input data into mathematical functions, called membership functions $\mu_A(x)$, $\mu_B(y)$, usually continuous, and these have the following characteristics:

- Core: part of the A set where there will be full membership.

    $$\mu_A(x) = 1, \mu_B(y) = 1$$

- Support: part of the A set where the membership is different from zero.

    $$\mu_A(x) > 0, \mu_B(y) > 0$$

- Boundaries: x elements of the set that

    $$0 < \mu_A(x) < 1$$
    $$0 < \mu_B(y) < 1$$

These values also specify the fuzziness level.

Crossover points of a membership function are all the x elements that assume a value equal to 0.5, $\mu_A(x) = 0.5$, $\mu_B(y) = 0.5$ (Figure 3.14).

The height of a fuzzy set is the maximum value of a membership function,

$$H(A) = \max\{\mu_A(X)\}$$

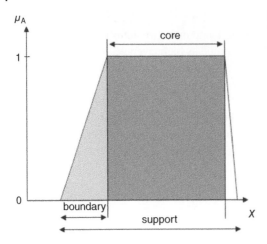

Figure 3.14 A membership function. Main definitions.

If H(A) <1, then the fuzzy set is subnormal. The height of a fuzzy system can be seen as the validity or credibility level of the information expressed by A or B. If A is a normal fuzzy set, but convex in one only point, then a fuzzy number is associated with it.

We cannot use mathematical functions to determine all that is found in nature, and to do this with a set of fuzzy elements takes us onto a tortuous path. Therefore, as for neural networks, we will try to use a figurative method to explain this. Every geometric shape can be described through a mathematical function. Membership functions are based on deductions or numbers. In literature, there are a lot of functions used to define fuzzy logic, and we will explain the simplest ones so we can unfold the notions at the basis of this theory. In the previous picture, we have a generic membership function shaped like a trapezoid, but we can also use sigmoid, Gaussian and triangle functions.

Once we have defined the membership function, we can establish the A sets (physical performance) and B sets (psychological performance) as follows:

$$A = \left(\frac{\mu(x)_1}{x_1} + \ldots + \frac{\mu(x)_n}{x_n} \right)$$

$$B = \left(\frac{\mu(y)_1}{y_1} + \ldots + \frac{\mu(y)_n}{y_n} \right)$$

A "normal" fuzzy set is one where the membership function has at least one x element with a membership value of 1. On the other hand, a set where only one element has a membership value equal to 1 is indicated as a set prototype.

A fuzzy set is convex if for x increasing values the membership function is monotonically increasing and then monotonically decreasing. A fuzzy set is not convex if for x increasing values the membership function is monotonically decreasing and then monotonically increasing. It is important to highlight that this definition of convex functions is different from the mathematical one. Generally, a function is convex when a line goes through two points over the diagram that represents it.

The main rules of fuzzy theory are defined to avoid a separation between the provided data and to express a measure that belongs to all sets considered in an application. Let us revise the following:

- Union Figure 3.15

$$A \cup B = \max \{\mu(x_i), \mu(y_i)\}$$

- Intersection Figure 3.16

$$A \cap B = \min \{\mu(x_i), \mu(y_i)\}$$

- Complement Figure 3.17

$$\overline{A} = 1 - \mu(x_i)$$

Figure 3.15 Union between two sets.

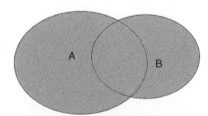

Figure 3.16 Intersection between two sets.

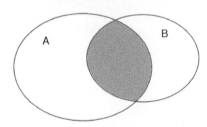

Figure 3.17 Complement of a set.

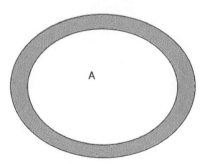

- Difference Figure 3.18

$$A \mid B = A \cap \overline{B} = \min \{\mu(x_i), 1 - \mu(y_i)\}$$

- De Morgan's Law Figures 3.19, 3.20

$$\overline{A \cap B} = \overline{A} \cup \overline{B} = \max \{1 - \mu(x_i), 1 - \mu(y_i)\}$$
$$\overline{A \cup B} = \overline{A} \cap \overline{B} = \min \{1 - \mu(x_i), 1 - \mu(y_i)\}$$

Once we have established the membership rules for input and output of the model, we can establish a few rules of syntax that could help define criteria if we know the C evaluation – for example:

1. if the physical performance is mediocre and the psychological one is sufficient, then the psychophysical status is low.
2. If the physical performance is inadequate and the psychological one is adequate, then the psychophysical status is medium.
3. If the physical performance is adequate and the psychological one is sufficient, then the psychophysical status is high.

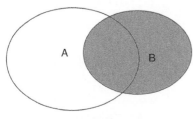

Figure 3.18   Difference between two sets.

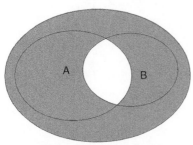

Figure 3.19   De Morgan's law: union between two complement sets.

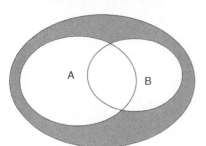

Figure 3.20   De Morgan's law: intersection between two complement sets.

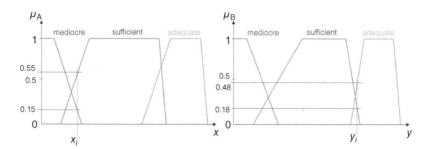

Figure 3.21   Fuzzy Logic typical rules.

4. If the physical performance is adequate or the psychological one is adequate, then the psychophysical status is medium.

Would you be able to convert these rules in mathematical form? What is important to remember, is how to mathematically use the logical operators AND and OR.

- AND

$$A \cap B = \min\{\mu(x_i), \mu(y_i)\}$$

- OR

$$A \cup B = \max\{\mu(x_i), \mu(y_i)\}$$

Let us delve into the membership function graph having x for physical performance and y for psychological performance. The input pair must be provided by the group of experts, while we only have to create a functioning model to establish a working general condition for the C set (Figure 3.21).

We apply the four rules above, obtaining the value of the activation function at the model output, i.e. the psychophysical status.

Rule 1 (Figure 3.22):
$A = \text{mediocre}, \mu_A = 0.15$
$B = \text{sufficient}, \mu_B = 0.18$
$\min \mu = 0.15$
$\rightarrow C = \text{low}$

Figure 3.22   Fuzzification – Rule 1 application.

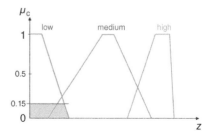

Rule 2 (Figure 3.23):
A = sufficient, $\mu_A = 0.55$
B = adequate, $\mu_B = 0.48$
min $\mu = 0.48$
$\rightarrow C$ = average

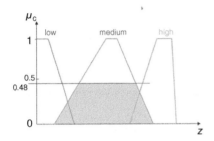

Figure 3.23 Fuzzification – Rule 2 application.

Rule 3 (Figure 3.24):
A = adequate, $\mu_A = 0.0$
B = sufficient, $\mu_B = 0.18$
min $\mu = 0.0$
$\rightarrow C$ = high

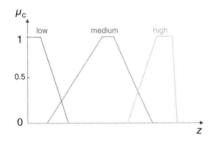

Figure 3.24 Fuzzification – Rule 3 application.

Rule 4 (Figure 3.25):
A = adequate, $\mu_A = 0.0$
B = adequate, $\mu_B = 0.48$
max $\mu = 0.48$
$\rightarrow C$ = medium

Figure 3.25 Fuzzification – Rule 4 application.

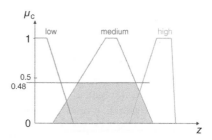

The sets we have obtained by intersecting the value of the outputs, obtained from each rule, will be joined together as a single membership function defined in the set C, from which we will obtain the value of the psychophysical status – for this purpose we can use the centroid method or the Sugeno's method (Figure 3.26).

$$z_i = \frac{\sum \mu_C(z) \cdot z}{\sum z}$$

The clearer our rules, the more reliable the model we can define. This implies that we can establish more criteria and combinations, by creating a linguistic rule based on the observed event and without having to conduct other math speculations. Fuzzy modeling is very practical and can be used with a limited amount of information. We can introduce some calculus algorithms as we have previously done for neural networks, such as the method of least square, the gradient one or the clustering one.

We can also evaluate the psychological performance or the psychophysical one only through the fuzzy method. If we are only evaluating the psychological performance, we must define a membership level for the athlete's mental health, and define the membership set by combining all the characteristics (trust, refusal, participation, sacrifice, motivation, and so on).

Figure 3.26 De-fuzzification – Center of Gravity (CoG) Method.

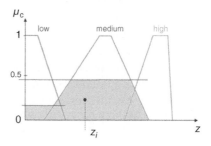

The example of the psychophysical level of an athlete helps us demonstrate that a system interested in the mental health of a human being is analyzed through fuzzy logic, as the amount of data cannot be acquired in a conventional or analytical way. A general but quick solution can be used at the beginning to save on computer costs, or in the variety of situations where the problem inputs are vague, dubious, or unknown.

Fuzzy logic is a method that formalizes the human ability to adapt to uncertain situations. We have noted how difficult it is to accurately model the psychophysical aspects by using a mathematical approach that is non-linear and unconventional – also limited to a knowledge base. The expert uses information and knowledge acquired through experiments to develop accurate models of the system's behavior.

3.11 It Is all in the Tools

Can we affirm that the complexity of an observed event comes from uncertainty? During the first years of modern civilization, we as human beings have faced issues of huge ambiguity that have challenged human thought. These ever-present characteristics are part of social, technological, and economic issues.

Why is it then that computer, created by human beings, are not able to tackle such ambiguous and divisive issues?

How can human beings' reason on real systems, when we are asking reality itself to give us more details than we could ever understand as human beings?

The answer is that human beings have the capacity to think in an approximate way, which is something that computers lack.

When thinking about a complex system, human beings only hold a general understanding of the issue. Luckily, this ambiguity and general understanding are sufficient for human beings to understand complex systems. Zadeh, father of fuzzy logic, used to say that the closer we look at an issue of the real world, the fuzzier the solution will be. It might sound as a paradox, but then how is it possible to find an exact solution if our approach is approximate?

As we learn more about a system, its complexity decreases and our understanding increases. As the complexity decreases, the precision offered by our calculus decreases the uncertainty encountered in the learning phase (does this remind you of anything? Neural networks). Logic only allows a general, not superficial, learning of the issue, and for overly complex systems in which very few numeric data or ambiguous information are available, the possibility of combining different methods will help our understanding of a general system, allowing us to mix the input and output data we have collected.

A tool, integrated in a complex system, means operating on data sources to convert the data into useful information or actions, can be related to the systems engineering approach.

Considering that some changes are needed to improve a system, the focus could be on the operational processes or also on the design processes. We take the perspective of using software to implement the requirements, functionality, and behaviors of a system. While systems engineering principles could certainly be applied to the development of these types of products, which provide specific services, functions, or features. This type is most easily recognizing with systems engineering in that an underlying principle has assigned functionality to specific subsystems, including subsystems software-based. In this section, it is shown that a generic tool can relate the various system processes. Every system development passes through a series of phases tied to the concept of a life cycle, which is defining for planning the activities, resources, schedules, and other supporting activities. A complex system development passes also for the decision-making process in order to keep the system objective on the fixed budget.

By the way, a system life cycle for developing and producing, consists of a series of systematic steps toward devising a technical approach; approach, engineering the hardware/software system; validating its performance; and producing as many units as required for distribution to the users/customers.

Support tools, such as software, can assist the PM or the project lead in developing computational models for the most delicate phase: decision making. The quality and availability of these tools determines success or failure. Support tools are used during the whole life cycle of the product, and this is why their development requires an important investment in terms of time and costs. This tool includes all the information on the decision-making process for a single system, to increase the efficiency and make the right decisions. It will be used to manage every aspect of the decisions – from allocating priority to the selection of the best supplier or candidate, to simply improving a process or product. We will see in the exercises that these tools can help us engage in new thinking.

A support tool must have scientific and calculus methods as well as guidelines. It also must feature abstract and practical techniques resulting from experience: through a heuristic approach, such as a general investigation that is needed to establish facts. This type of research method, not based on mathematical algorithms, allows us to foresee and make plausible a result, which will then need to be checked and validated formally. The quantitative and systematic (or systemic) approach is essential, but the search of all the qualitative aspects must be carried out through heuristic approaches.

The sports club example helps us define the essential aspects for a possible support tool, which is included in the general architecture of the system or subsystem

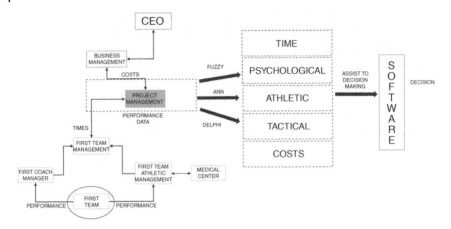

Figure 3.27 From sub-system architecture to software development.

of interest, as shown in Figure 3.27. The architecture is surely an important guide toward finding a valid decision-making support. Such support requests are usually: reliability, long term durability, coherence with the stakeholders' requirements.

Finally, we should not forget that, in the initial phase, the development of a "new" system or sub-system (existing system or sub-system with ANN) requires a complex effort extended up to the in-service phase. A first phase of development concerns the concept of the "new" system, in fact this phase can be featured by the "new" system requirements associated with the stakeholder's need. Therefore, the "new" concept (existing system or sub-system with ANN) is defined by comparing and verifying all the different correlations that exist between a heuristic approach and a physical–mathematical one.

3.12 Chapter Summary

Systematic and systemic approaches enable the interpretation of reality. In the case of animals, we need to find a link between zoology and IT, statistics and veterinary medicine. We will rely on systems engineering to help us manage the complexity of systems as a whole.

Systems engineering has been used to solve complex problems since the 1950s. It was formally recognized by the founding of the INCOSE in 1994. AI and robotics are powerful tools at our disposal to create simple and complex systems from our children's home to constellations of spacecraft. ANN have become increasingly important in systems engineering. Complexity is an intrinsic characteristic of a system, but also a regulating factor of its life cycle.

The approach of systems engineering will highlight the concepts of "system elements" and "system interfaces."

Making a decision is not as simple as it may seem due to the many variables at play. An optimized decision-making process should assess the reliability of the primary components of the aircraft. A computational decision-making process is a far-reaching subject that uses math to evaluate a big amount of data and determine the most convenient solution or line of action. The aim is to optimize the main efficiency parameters, all the while supporting the management of the project. System engineering allows to analyze each sub-system independently of the other ones. Analytical techniques alone are not sufficient to define the different levels of a system.

PM is a key figure in the implementation of any project. Their main goal is to harmonize costs, time frames, quality and client satisfaction. The PM's activities can be divided into two main areas – operational and relational. They gather technical data (player performance, business information, profit, costs, investments).

Systems engineering requires a knowledge of the elements to be analyzed, with which one can develop a process aimed at performance improvement. In this sketch, we will define the main parameters required to create a mathematical model, which will help the decision-making process carried out by the sports manager.

The introduction of neural networks provides a valid technical support. If correctly integrated, they can be managed with a "*V*" diagram i.e. a mathematical model of a neural network.

Questions

1 Why is it necessary to talk about a system of interest when a neural network is introduced?

2 Can all the main processes of the SE be adapted to neural networks? What can the compatible processes be for Neural Networks? What are some practical applications?

3 How can a Neural Network be used in a decision making process?

4 Why can other methods (e.g. fuzzy) be integrated with neural networks?

5 Can a Neural Network directly solve a problem of decision making without going through the main SE processes during the whole system life cycle?

Sources

Ross, J. (2010). Timothy, *Fuzzy Logic with Engineering Applications*, 3e. Wiley.

Maier, M.W., Rechtin, E. (2000). *The Art of System Architecture*, 2e. CRC Press.

Denardo. (2001). *The Science of Decision Making: A problem-based approach using Excel.* Wiley.

(2015). *INCOSE Systems Engineering Handbook: A Guide for System Life Cycle Processes and Activities*, 4e.

Engel, Robert (1965) *Machol System Engineering Handbook*, McGraw-Hill.

Chestnut, Harold (1967) *Systems Engineering Methods*, Wiley.

Blanchard, Benjamin S. and Wolter J. Fabrycky (1981) *Systems Engineering and Analysis*, Prentice-Hall.

Part II

Neural Networks in Action

4

Systems Thinking for Software Development

[…] When an algorithm or effective procedure is used to calculate the values of a numerical function then the function in question is described by phrases such as effectively calculable, or algorithmically computable, or effectively computable, or just computable […].

Cutland N.J. (1980).
Computability: An Introduction to Recursive Function Theory.
Cambridge University Press.

Some computational methods[29] have been mentioned in this book which were developed over the past 50 years to achieve prediction, classification, and clustering challenges. We will build a computational method that has been found to be successful in a wide range of social, engineering and sports businesses.

We have mentioned the definition of system in the previous chapter which will provide an important guideline for the creation of analytical software. The objective of the engineering development phase of a "new" system is to integrate the AI/ANN, following the approved "feasibility study" and to evaluate (verification and validation) the functionality of the "new" system.

What is the purpose of the software? In what context should it be inserted? Moving beyond the analytical formulation we need a new language to be able to handle the complexity and large amounts of data required to resolve machine learning problems, stored in extensive multidimensional matrices. As shown in Figure 4.1, a working group collects and arranges data in an appropriate repository. Simply collecting data is not enough, if the data cannot be managed by a database. Messy or raw data is not meaningful for a neural network. Therefore, big data must be well sorted, cleaned and validated, following a verification phase, as it must be ready for software immediately. A handling source data shall be complied with all techniques of neural network software.

In general, computational methods such as those of neural networks, are based on the concepts expressed in the previous chapters and can also be improved

Systems Engineering Neural Networks, First Edition. Alessandro Migliaccio and Giovanni Iannone.
© 2023 John Wiley & Sons, Inc. Published 2023 by John Wiley & Sons, Inc.

Figure 4.1 The challenge of handling big data. Source: Image by Peggy und Marco Lachmann-Anke from Pixabay.

by the quality of dataset used. Let us start early in saying that feasibility studies provide an excellent opportunity to manage stakeholder expectations and risk: from the onset, managers should understand that the project may not actually be feasible, or the details may change based on data and context. It has been seen that designing a system requires a thorough study of technical feasibility. For "first generation" software, almost anything always seems feasible. Modern microprocessors and memory chips can accommodate large software systems. Limits on size, strength, or accuracy are not contemplated for such software unlike the limits imposed for hardware components. In general, technical feasibility for software tends to be taken for granted. This is a great advantage even for "next generation" software, but it invites the developer to take on challenging requirements. However, the resulting complexity proves too difficult and expensive. During the feasibility study phase, one must ask how quickly the intelligent system can get to a baseline model performance for the task and how quickly it can show measurable progress on that baseline. The AI/ANN (AI focused on Neural Networks) software development is feasible by leveraging the environmental simulation and distributed architecture capabilities of a complex system. Modeling is a capability that can be added to develop a complete model useful for simulating and studying the problem and designing the subsequent transitions from simulation to real-world software and operational hardware. As we have mentioned in previous chapters, an AI Software can iteratively cycle through experiments using different model architectures and parameters; it could process hundreds of variables. This means that you must build the best architecture for the task; additionally, you must figure out if there is a solution to the problem and if that solution can provide measurable results or a measurable increase over the baseline. Thoughtful choices about experimental design are to be made before the iteration process begins.

The complex nature of a system states that the transition from simulation to a demonstration exercise would be best accomplished by first implementing real-time AI/ANN software code and refining the technology in an interactive learning process. The feasibility of an intelligent system is considered a Multi Agent System proposed in the literature that initially offers specific functions. A comprehensive analysis is conducted regarding the investment cost, as well as the profits and benefits. Any intelligent system must provide additional services to its initial purpose to maximize the benefits of its implementation. The software can be constantly updated with a simple heuristic algorithm regarding loss reduction. The initial investment cost of implementation must be compared to the combined profits and benefits that the system will produce for its stakeholders.

Below we list the key points of a feasibility study:

- Open-source datasets.
- Complex system experience.
- Identification of an optimal analytical architecture for data collection, by Machine Learning techniques and/or heuristic algorithms.
- Categorization of available data for further reuse.
- Analysis and observation of events.
- Identification of a data management system
- Creation of a data communication viewer.
- Creation of a dashboard design for decision making.
- Support for institutional communication of available statistics and databases.
- Software implementation times.
- Comprehensive investment cost analysis.
- Software management needs to constantly share results back and forth between the business and technical teams. Such management serves to create direct communication between these areas of interest.

When improving an existing system, in contrast, the developer often directly addresses the task, i.e. when an algorithm is already in place. The task is to make a measurable enough improvement on this basic configuration such that it merges to the existing production. This implies that the timing of the initial explorations, once provided, is also acceptable.

For example, if the potential delta in business impact is very high, that time can be quite long, sometimes up to 6–12 months. If the potential delta in business impact is lower, it is better to keep the time shorter and then experiment quickly with newer algorithms.

Another noteworthy nuance occurs when the AI software developer uses off-the-shelf trained models that can tune to the specifics of the domain and environment. With accessible pre-trained models, the developer already knows that the solution will work; what they are trying to find out now is how well the models fit a specific domain.

4.1 Programming Languages

The choice of programming language is one of the major decisions in software design. Critically dependent on the type of system, whether military or commercial, whether real-time or interactive, it follows that the purpose of its application should take priority. A language can impact maintainability, portability, readability, and a variety of other characteristics of a software product. Programming languages are usually coupled with a database system and are tied to the use of structured query language (SQL). A key feature of languages is to bring the software programming environment as close as possible to the natural language domain and to provide interactive tools to create solutions. For example, creating a user input form on a workstation is done interactively with the programmer. The programmer inserts labels and identifies allowable input values and any restrictions.

Some of the most known programming solutions used in this book are;

- Visual Basic (VB): A language that allows graphical manipulation of subprogram Objects and Subroutines, Graphical applications and User interfaces
- Python: A powerful, general – purpose language that implements object – oriented constructs Objects, Functions, Simulations, and Real-time applications.

Some of the examples in Chapter 5 are programmed with Microsoft® Excel® 365 (related to VB). Furthermore, given the popularity and huge number of online materials on how to install and use this software, no further guidance will be provided in this book. Examples 5 and 6 examples are developed by the Python3 programming language which can be used in two ways:
Online:

- Using Google Colab, the files can be requested from AiShed (http://www-ai-shed.com)

Offline:

- By adapting the code for local execution by installing PyCharm 2020 or similar compiling software

Microsoft Excel (MS Excel)

Figure 4.2 NN structure for reference.

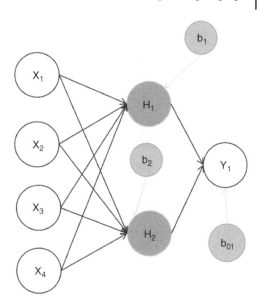

Excel is a powerful tool for creating practical mini-tools and starting the first steps in the Neural Networks. For instance, let us see how a worksheet can be set up to calculate the variables of a cost function.

Before proceeding, it is important to understand how to calculate matrix[41] scalar products.

Let us define a neural network consisting of four input neurons, two neurons of a single hidden layer and a single output neuron (See Figure 4.2).

Thus, two weights (W) are considered for each input neuron, for a total of eight, and a single weight for each neuron in the hidden layer, for a total of two. As shown in Figure 4.3, a model of NN structure is defined using possible syntactic rules of the programming language considered in the following numerical example.

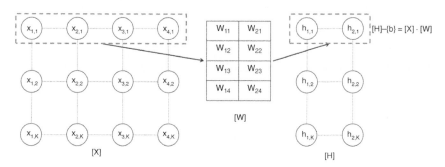

Figure 4.3 Model for calculation – Input Layer.

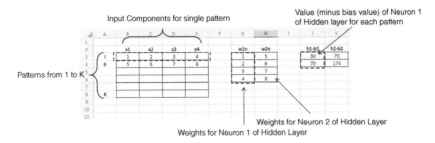

Figure 4.4 Neural networks nitty and gritty.

Let to fix any spreadsheet data as per Figure 4.4.

The matrix product[37] can be computed from two matrices, therefore select the cells representing values of the hidden neurons and use the formula as shown in the Figure 4.5. Remember: finding the product of two matrices is only possible if the inner dimensions are the same!

In such case, the input values are defined as a vector (list or mono-dimensional array of numbers) for each pattern, therefore a matrix is considered to include all the available patterns. The other matrix is filled with the values of weights.

A scalar product (See Figure 4.6) is found for each neuron of the hidden layer using the input vector of a single pattern and the weight vector for a single neuron of the hidden layer.

Note that the scalar product may not work if the two vectors to be multiplied are defined in a row and a column respectively. So, the vector of weights was transposed and written on a row like the input vector.

	A	B				F	G	H	I		J	K
			MATR.PRODOTTO (matrice1; matrice2)									
1												
2			x1	x2	x3	x4		w1n	w2n		h1-b1	h2-b2
3		I	1	2	3	4		1	=MMULT(B3:E4;G3:H6)			
4		II	5	6	7	8		2	6		70	174
5								3	7			
6								4	8			
7												
8		K										

Figure 4.5 MMULT function (or the equivalent in your language) – Matrix Product.

	A	B					H	I	J	K	L	M	N
12													
13			x1	x2	x3	x4		w11	w12	w13	w14	h1-b1	h2-b2
14			1	2	3	4		1	2	3	=SUMPRODUCT(B14:E14;G14:J14)		
15								w21	w22	w23	w24		
16								5	6	7	8		
17													
18													

Figure 4.6 SUMPRODUCT – Scalar Product.

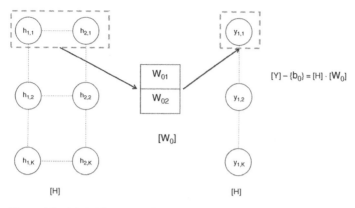

Figure 4.7 Model for calculation – Hidden Layer.

Select the cell representing the value of a hidden neuron and set the formula as shown below.

The scalar product, in fact, must be defined for each cell, while the matrix product is calculated automatically by the program, once the formula has been written in a single cell.

Considering the model for calculation shown in the Figure 4.7, we want to explain that the biases are scalar values and they can be estimated with the outputs as necessary. Following the example, there are three biases in total, two for the hidden layer (See Figure 4.7), and only one for the output neuron (See Figure 4.8). A matrix product is applied to provide the weights array (See Figure 4.9), multiplying the hidden neurons values with combination between hidden layer bias and output values.

The activation functions and the value of temperature (See Chapter 6) of each single neuron have not been introduced in the spreadsheet. We will leave this exercise to the reader.

It is also possible to use ready-made feature such as the Excel *Solver*, which can be easily installed as an add-on to worksheet, and it will be used to solve some of the exercises covered in this book. In this application, all the concepts expressed

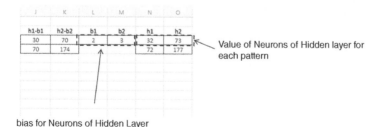

Figure 4.8 Fill the spreadsheet.

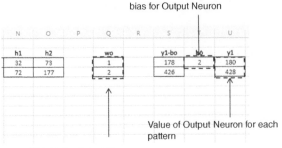

Figure 4.9 Weight, Bias and Output NN in the spreadsheet.

in the next chapters are shown, such as iterative methods and optimization of a cost function.

In particular, the cost function is written in a cell in which a formula is defined, and the formula can be changed if it is more appropriate for our purposes. The cell takes reference of the values written in other cells (weights, bias, probabilities), which will become the variables of our model. (See Figure 4.10)

This is the procedure to be applied, following the instructions contained in the: Figure 4.11

(1) Enter the reference of the cell with the cost function or a value to be optimized.

(2) Set the value to be assumed by the cost function: MAX, MIN or Exact Value.

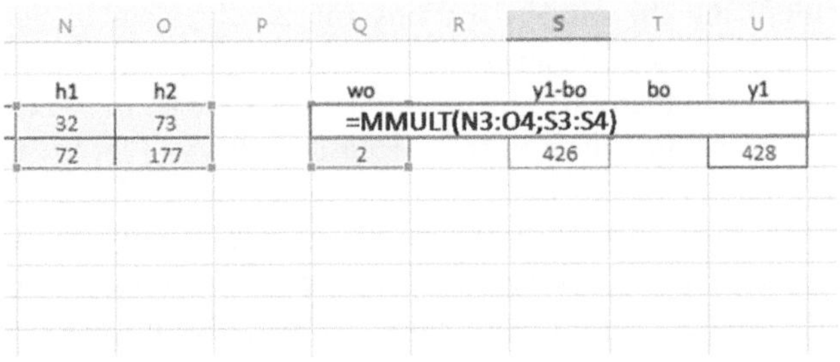

N	O	P	Q	R	S	T	U
h1	h2		wo		y1-bo	bo	y1
32	73		=MMULT(N3:O4;S3:S4)				
72	177		2		426		428

Figure 4.10 MMULT – Matrix Product.

Figure 4.11 Set Objective to Maximize, Minimize or exact Value.

Figure 4.12 Select the variable cells.

Figure 4.12

(3) Enter the reference of the cell or cells containing the variables of the cost function (weights and bias).

Figure 4.13

(4) Enter the reference of the cell in which a value or a function is to be limited according to the available relationships: equal, less-than-equal, greater-than-equal, only integers, only binary numbers or none of the previous relationships.

(5) Enter the limit value for the cell defined in the point 4. This cell is active only for the relationships: equal, less-than-equal and greater-than-equal.

Once the cost function and any constraints have been defined, choose the iterative method from the following options:

- Nonlinear LSGRG to optimize nonlinear and differentiable functions
- LP/Simplex standard to optimize non-linear functions.
- Evolutionary to optimize non-linear and non-derivable functions.

Note: The evolutionary method is not explained at this point since we will only deal with differentiable functions.

For the LSGRG method, used in the exercises 1, 2, 3 of chapter 5, see Figure 4.14:

Figure 4.13 Define the constraints.

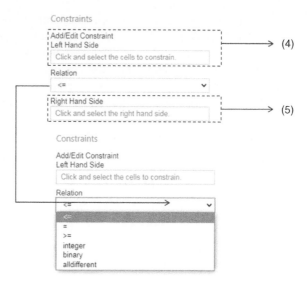

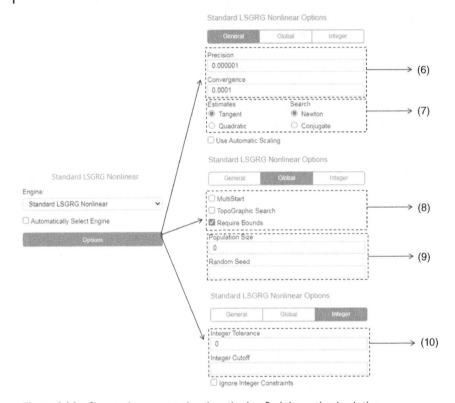

Figure 4.14 Choose the computational method to find the optimal solution.

(6) Set a tolerance value for any stop of the iterative process and set a convergence value to which the subsequent iterations must approach.

(7) Choose the derivative method between tangent and quadratic, and the iterative method between Newton and Conjugate.

(8) Select type of research for stationary points of the cost function.

(9) Write an initial value of the random number generator to find stationary points of the function.

(10) Define any limits on integers found in the function.

For the Standard LP/Simplex method, see Figure 4.15:

(11) Define the limits on decimal or integer numbers found during optimization of the imposed value.

(12) Choose type of feature of the iterative method: automatic, aggressive or neither.

(13) Choose iteration criteria, whether to impose an automatic scaling factor between one iteration and another, and any constraints on integers.

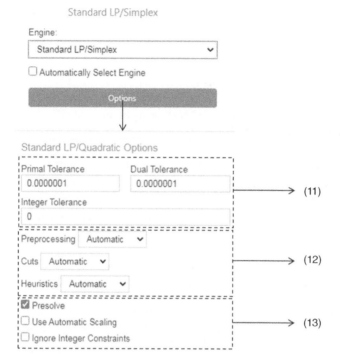

Figure 4.15   Options of a computational method.

Programming in VB can be time efficient so as defining the spreadsheet automatically before entering data which can be is a key advantage for large datasets.

Google Colab

In this book we will use Google Colab to simplify execution of examples written in Python 3. A Colab notebook is nothing more than an interactive document with executable code, images and text which we will use to explain the examples step by step. For more information on Google Colaboratory, you can consult the sources section on the blog connected to this book.

PyCharm 2020.1.4 (Community Edition)

Another way to run examples on a local PC is by using an integrated development environment (IDE), such as PyCharm.

PyCharm: How do you install PyCharm?

To get started, download and install the community version of PyCharm:

- Download for Mac (open the downloaded .dmg file and drag PyCharm to the Applications folder)
- Download for Windows (Open the downloaded .exe file and install PyCharm, using all the default options.)

Table 4.1 Python libraries for coding a neural network.

Numpy

A library for the Python programming language, adding support for large multi-dimensional arrays and matrices, along with a large collection of high-level mathematical functions to operate on these data structures.

Pandas

A software library written for the Python programming language for data manipulation and analysis. It offers data structure wrapper and operations for manipulating numerical tables and time series.

Matplotlib

A plotting library for the Python programming language and facilitates NumPy. It provides an object-oriented API for embedding plots into applications using general-purpose GUI toolkits like Tkinter, wxPython, Qt, or GTK+.

Sklearn

A software machine learning library for the Python programming language. It features various classification, regression and clustering algorithms including support vector machines, random forests, gradient boosting, k-means and DBSCAN, and it is designed to interoperate with Python scientific libraries NumPy and SciPy.

Itertools

The module standardizes a core set of fast, memory efficient tools that are useful standalone or in combination with other libraries. Together, they form an "iterator algebra" which enables construction of specialized tools succinctly and efficiently in pure Python.

Keras

An open-source library that provides a Python interface for artificial neural networks. Keras acts as an interface for the TensorFlow library.

Given the volatility of the software, we will use default installation guidelines, as they may already be obsolete by the time the book is published. Hence, we will refer to the blog connected to the book for updated information on installing the Python development environment that will best suit our needs.

PyCharm: Libraries

Python makes use of libraries/packages/modules as a source of functions that are pre-programmed and ready to be called up by the user when necessary to save time and system resources.

The following Table 4.1 will be required for the examples in this book some useful libraries (source Wikipedia):

Note: To install the libraries on PyCharm

PyCharm provides methods for installing, uninstalling and updating Python packages for a particular Python interpreter. The blog connected to the book has useful links on installing the necessary packages.

4.2 One More Thing: Software Engineering

When working on a software engineering team, the terms software engineering and software systems engineering, are sometimes used synonymously. Such difference may not be immediately recognized especially within a discipline that is largely based on computational science, so it is essential that roles are effectively defined when AI outcomes are to be achieved using a SE approach. The former refers to the development and delivery of manufactured, stand-alone or embedded software. The second refers to the application of principles to the discipline of software engineering. Many principles, techniques, and tools are similar for both fields, and research has facilitated the evolving merger. The software improvement in a complex system is an integrated and comprehensive part of the system development. Thus, systems engineering must include software engineering as an integral discipline, not simply as another design effort to implement functionality. This versatility and potential power make software an indispensable ingredient in modern systems, both simple and complex.

An AI-based software development process (SDP) can be resolved into the following steps:

- system analysis including system objective analysis;
- software requirements;
- software architecture;
- software design;
- programming (development/implementation)
- software design procedures (documentations);

- software test (formalize the task of developing);
- integration in the system.

For the AI software, the popular life cycle models can be again used. The various software models involve the same basic functions, differing mainly in the manner in which the steps are carried out, the sequencing of activities, and in some cases the form in which they are represented. Overall, AI SDP can generally fall into four typical categories:

- **Linear:** During the life cycle, the steps of the process are sequential, and each step has feedback. Linear development works well when the object is clear and consistent with stable requirements, reasonable schedules and resources.
- **Incremental**: uses the same basic steps as linear models, but each step can be repeated in multiple iterations. Also, each step can be performed with different degree of detail than the others. They work well in stable requirements environments where partial functionality is desired before the complete system is developed.
- **Evolutionary:** is like the incremental concept but works well in environments where the features and attributes of the final product are not known at the beginning of the development process. Evolutionary models provide limited functionality, they can apply in environments for experimentation, demonstration and familiarization. For evolutionary models, it is essential to constantly receive feedback if the system changes to meet user needs.
- **Agile:** deviates more from the aforementioned three categories. With linear, incremental, and evolutionary models, steps are manipulated in different sequences and are repeated in different ways. In agile development environments, the steps are combined in some way and the boundaries between them are lost. Agile methods are appropriate when environment change is the constant throughout the process.

An AI software must be considered in accordance with other all system elements, such as the following:

- Physical entities whose function and operation are being monitored or controlled.
- Actuators that receive an action from the system (e.g. Robot).
- Data transmission to other computers.
- Humans who will interact with the system.

The AI/ANN software elements may appear complex in their functions and may be critical to the proper functioning of the system. Therefore, to prevent the system from becoming unstable and difficult to manage, special risk mitigation measures must be considered. The risks of a complex system may be related to the computer sub-system, they must go through the application of a systematic and logical process of software identification, analysis and control. Therefore, further analysis is required to identify and assess the risks associated with the use of the software in order to rule out a good chunk of the entire system errors. In the following, any AI Software Errors are listed:

- **Incorrect algorithms:** the software may perform calculations incorrectly because of mistaken requirements or inaccurate coding. Algorithms produce incorrect or no output data; or both.
- **Data errors**: the software may receive out of range or incorrect input data, as such as wrong data type or size.
- **Large data rates**: the software may be unable to handle large amounts of data or many user inputs simultaneously.
- **Logic Errors**: the software may receive bad data but continue to run a process, thereby doing the right thing under the wrong circumstances.
- **Interface errors**: a message may be incorrect or unclear, leading to the system operator making a wrong decision.
- **Poor interface design and layout**: an unclear graphical user interface can lead to an user making a poor decision.
- **Server overwriting:** Improper memory management may cause overwriting of memory space and unexpected results.
- **Multiple events occurring simultaneously**: a system operator may provide input in addition to expected automated inputs during software processing.

As you progress through the feasibility study phase and explore the performance and architecture of the base model, you may find that you need to reformulate the task. The business team must then decide if that scaled-down version will add enough value, and if so, that version may end up being the first Minimum Viable Product (MVP) that is developed further in the future. If the business team decides that the scaled-down version does not provide enough value, however, it should probably be stopped so that efforts can be prioritized toward another task. As these discussions occur, product managers will want to take the time to work with the design to understand the cost of errors in the production solution. Essentially, you want to know how often the system needs to be corrected to be useful. The acceptable error rate may change: a product recommendation engine, for example, will certainly have a different acceptable error rate than a banking or healthcare solution.

Unlike traditional development workflows, the AI/ANN software feasibility study is used to dig into the data and quickly conduct experiments to establish baseline performance on a task. And it requires project managers to iteratively reassess impact and feasibility as teams learn more about the problem space and possibilities. The recommended order of precedence for eliminating or reducing risk in the use of AI software and computing system follows:

- Design for minimum risk.
- Incorporate safety devices.
- Provide warning devices.
- Develop and implement procedures and training.

Finally, a risk mitigation can be also estimated during the operational phases. Not all analysis methods used to identify software and compute risks are covered in this chapter. A complex and integrated system comprised of hardware, software, human and environmental interactions, are included in the risk reduction process. The effort to produce software should take into account the fault avoidance, removal, detection, and tolerance within the constraints of operational effectiveness, time, and resources throughout all phases of the life cycle of a system. For an effective software, a developer should consider using a combination of these aspects: computing system development and computing system risks.

An operator should prepare a Software Development Plan (SDP). Upon developing valid requirements as a result of efforts to identify, characterize, and reduce the risks and assuring the integrity of the software and proper implementation of the safety requirements, the SDP should describe the activities, methods, and standards for the development of safety-critical software to reduce the software risks and assure software integrity with respect to safety. The SDP should include both management and engineering of the software safety effort. In addition, the following topics should be taken into account:

- Purpose, scope, and objectives of the software safety program and its tasks.
- Organization and responsibilities.
- Schedule and critical milestones.
- Staff training requirements specific to the software design, development, testing, implementation, and maintenance.
- Contract management.
- Tools approved for usage, such as computer-aided software and simulators.
- Design, coding, and safety standards and guidance documents.
- Software System Safety Engineering
- Methods for identifying safety-critical computer system functions.
- Methods for performing software and computing system hazard analyses that generate software safety requirements.
- Approaches to validation and verification testing, analyses, and inspections.
- Software configuration management.
- Software quality assurance.
- Installation processes and procedures.

- Maintenance activities.
- Anomaly reporting, tracking, root cause analysis, and corrective action processes.
- Training requirements.

Once the planning is completed, an operator should identify and describe safety-critical computer system functions. The operator would normally do this in three steps:

- Identify risk-critical system functions.
- Describe risk-critical system functions.
- Identify risk-critical computer sub-system functions.
- Describe risk-critical computer sub-system functions.
- Relate risk-critical computer sub-system functions to risk-critical system functions.
- Identifying those functions that are essential to operating without risk is the first imperative.

As we have repeatedly mentioned, the risks of system, considered in both the design and in-service phases, must be promptly identified, analyzed, according to the likelihood of impact, and consequently managed (solve risks and/or mitigate risks).

Based on the importance of adopting well-recognized software engineering processes to manage any type of risk, associated with:

- ANN software,
- existing system (i.e. sports club, aircraft, photovoltaic system)
- "new" system (existing system with ANN),

the systems engineer has to take part in the software development in order to support the risk analysis by system engineering methodologies. After describing the software functions (i.e. ANN software), the software engineer and the software developer should define and communicate the critical ones including those associated with the system according to the system's requirements. Examples include, but are not limited to, the following:

- Interfaces between the software and other sub-systems.
- Diagrams showing the data flow.
- Data transmission times.
- Power supplies for each software function.
- Sub routines to detect faults in the software.
- Firewalls for security.
- User display.
- Operator manuals and technical documentation.

It is possible that some functions will not be critical during some operational phases. In these cases, the user may monitor, control, and follow up with some guidance for non-critical functions. The system engineer must ensure that each function of the software meets the requirements of the system and complies with the objective.

Following the above functions, an application to the club sport context is consistent. Let us try to analyze in broad strokes what we have described and leave a more in-depth analysis to the reader. Introducing an AI/ANN software, the existing system risks can be reduced by associating its specific functions with them that will result in certain software functions. Isolating the elements of risk and decreasing the costs of a sports club would lead to a thorough analysis of all risk functions of the complex system.

The systems engineer needs helps to determine priorities within the safety effort, focus the use of resources, and adjust activities based on the most important risk concerns. Examples of safety-critical functions of a sports club system include, but are not limited to, the following:

- Profit statement.
- Residual profits.
- Communications.
- Merchandising.
- Stock exchange listing. The disappointing stock market performance cannot be improved just by winning more.
- Financial instability deriving from sports results.
- The high players value destabilizes the balance sheet.
- Non-diversification of extra-seasonal revenue sources as such as complementary services to show business (e.g. managing an owned stadium).

In summary, if we decide to develop a Neural Network (NN), a preliminary analysis phase is essential. A Neural Network can be designed to solve the issues of an existing system or sub-system, hence it is necessary to determine a proper functionality, availability, maintainability and reliability.

As stated earlier, a NN affects all information acquired from human understanding and consciousness as an outcome of learning about an actual event. Based on the human learning ability, embedded in a specific experience and settled in the environment, there is an opportunity to analyze the issues that are inherent in the existing system in which to put in a NN.

When we want to analyze the reliability of a NN, the system engineer must understand all the processes of the existing system to infer its reliability otherwise the data derived from it may be unusable.

A reliability analysis of NN software can also support the reliability analysis of the related sub-system.

Proper analysis is generally poor and limited by lack of appropriate data and understanding of interactions with the environment, humans, and other software.

Therefore, the software engineer and expert judgments are critical for NN software. Following the engineering development phase, we need to consider all the life cycle processes and methodologies (SE approach), including decision-making, risk management and configuration management. Finally, a preliminary development is deemed applicable on ANN design (software/application), defined as configuration item of "new" system or sub-system.

4.3 Chapter Summary

Take outs from this chapter:

- Some computational methods were developed over the past 50 years to achieve prediction, classification, and clustering tasks.
- Systems Engineering provides an important guideline for the creation of analytical software.
- It has been seen that designing a system requires a thorough study of technical feasibility, which tends to be taken for granted.
- The AI software development is made feasible by leveraging the environmental simulation and distributed architecture capabilities of a complex system.
- The choice of programming language is one of the major decisions in software design. A language can impact maintainability, portability, readability, and a variety of other characteristics of a software product.
- A key feature of languages is to bring the software programming environment as close as possible to the natural language domain and to provide interactive tools to create solutions.
- Some of the most known programming solutions are Visual Basic, often seen in connection with Microsoft Excel and Python, which can be used on a variety of platforms including PyCharm.
- When working on a software engineering team, the terms software engineering and software systems engineering, are sometimes used synonymously.
- Incorrect algorithms: the software may perform calculations incorrectly because of mistaken requirements or inaccurate coding.
- Product managers will want to take the time to work with the design to understand the cost of errors in the production solution.
- "Design for minimum risk" is recommended for eliminating or reducing risk in the use of AI software and computing system.

Questions

1 What are the differences between software engineering and software systems engineering?

2 What is meant by SDP?

3 How can an AI SDP be introduced into the SE standards?

4 What are typical AI software errors?

5 Why is Software System Management (SSM) included in the SDP?

6 What are the benefits of SSM in the NN applications?

Source

Kossiakoff, A., Sweet, W.N., Seymour, S. J., and Biemer, S.M. (2011). *Systems Engineering Principles and Practice*, 2e. Wiley.

5

Practice Makes Perfect

[…]The perception and definition of a particular system, its architecture and its system elements depend on an observer's interests and responsibilities.[…]

ISO/IEC 15288

In this chapter nine interesting examples will be presented, some developed with the MS Excel® spreadsheet and others with Python programming language. These exercises can be useful for not only understanding the algorithms based on mathematical models[30] for decision-making but also to provide a guideline for other applications for various needs. Therefore, it will be shown that neural networks are useful for problem solving, processing approximate data and providing consistent results of the examined event. Furthermore, we will see that neural networks can learn with a certain level of uncertainty, which can be already specified in the input data set.

The supervised learnings will be explained when the neural network solves a mathematical function, characterized by stationary points (let us only conform to the minima and maxima). Furthermore, such functions will allow us to reach or correctly set the optimum point of our model. If we believe that neural networks can provide a significant advantage when the cost function is introduced, it is important to explain that the cost function is an essential element for the optimization process in Linear Programming (LP). Considering that one of the main managers' objective is to allocate resources among activities or projects. LP is a widely used method of allocating resources optimally during design or operational activities. It is also successfully used as the beginning of decision making in almost all organizations. In particular, exercise 7 summarizes the concepts of cost functions in linear and non-linear programming by applying game theory or in general a *"simplex"* method.[16] It will be easy to deduce that in any model we want to build to facilitate the decision making process, we must define and study the mathematical function that will allow us to optimize our systems engineering processes.

Systems Engineering Neural Networks, First Edition. Alessandro Migliaccio and Giovanni Iannone.
© 2023 John Wiley & Sons, Inc. Published 2023 by John Wiley & Sons, Inc.

All the examples are derived from our professional experience or academic research. All the scenarios presented are measurable and, in the case of prediction, they are a starting point. The concept of forecasting will not be explained as a precise and absolute discovery but rather a reliable and plausible estimate. Reaching a goal means identifying all the boundary conditions and aligning to them. In other words, there exists no physical phenomenon that can be described without understanding its characteristics.

How can you explain lightning without even knowing the concepts of power, electricity and meteorology?

By now, you will surely have understood that the support of a mathematical model that can facilitate our reasoning and our decision-making abilities will require a certain amount of information. A large amount of information must always agree with model limits, or the optimal solution will be not achievable otherwise.

5.1 Example 1: Cosine Function

Verification confirmation, through the provision of objective evidence, that specified requirements have been fulfilled [ISO 9000 : 2000].

Validation confirmation, through the provision of objective evidence, that the requirements for a specific intended use or application have been fulfilled [ISO 9000 : 2000].

The purpose of this exercise is to show that the verification and validation processes, defined by the SE approach, are satisfied through the NN usage. In general, the NN verifies and confirms that a system element or a system has been correctly designed. In addition, the NN usage can conveniently reduce some verification and validation activities, correcting pre-emptively the design errors before completing of system development. The system validation is subject to approval by an authority and/or the stakeholders. This process is invoked during the Stakeholder Requirements definition process to confirm that the requirements correctly reflect stakeholder needs. The use of a NN establishes certain validation methods during the design phase by including system-level performance development over the entire operational regime. The system design team uses the results of this activity to predict success in meeting user and buyer expectations, as well as to provide feedback to identify and correct performance deficiencies prior to implementation.

Objective:

- Design a model that processes the cosine function.

Requirements:

Let us assume to be in need to design a system that approximate the value of the cosine function within a certain margin of error.

- The system shall be able to calculate that:
  - $\cos(0°) = 1$
  - $\cos(90°) = 0$
  - $\cos(270°) = 0$
  - $\cos(360°) = 1$
- The system shall be able to verify that cosine values:
  - decrease between 0° and 90° in the positive co-domain;
  - decrease between 90° and 180° in the negative co-domain;
  - increase between 180° and 270° in the negative co-domain;
  - increase between 270° and 360° in the positive co-domain.
- The system shall be able to calculate the cosine function in the range [0°, 360°].

Now, let us consider a cosine function which, together with a sine function, are the cornerstone of mathematics. Before the arrival of logarithms, the trigonometric functions were used to simplify calculations, and without them, it would not be possible to talk about kinematics and/or dynamics today.

Furthermore, it is possible to define a mathematical function associating numerical values to a set of qualitative evaluations through Fuzzy Logic.

In this exercise the function is known to apply a neural networks approach and the challenge would be effective training of the model.

Represent the following function graphically (see Figure 5.1).

$$y(x) = \cos(x)$$

The importance of normalizing input and output values has been highlighted such that all values are equally treated by the neural network model. The cosine function lends itself well to this "stratagem," as shown in the following Figure 5.2.

Once x-axis values have been normalized, the curve will be defined around its average value in the positive co-domain (y-axis). These steps will decrease calculation times during optimization process; in fact, the method will search for the best solution in a very narrow range (look at the derivatives of the activation functions and the iterative method).

Table 5.1 lists main characteristics of the neural network which is built to define the cosine function. In the Figure 5.3 the NN model is represented.

The normalized inputs are defined which are the values along the x-axis (degrees). There will be 60 and 14 data points for training and validation sets respectively. The y values of the cosine function are set as the target of the network, which are also normalized as much as possible by its average:

$$y = 0.15 + Norm[\cos(x) + 1] \cdot 0.7$$

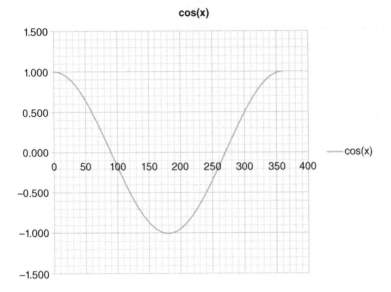

Figure 5.1 Cosine function.

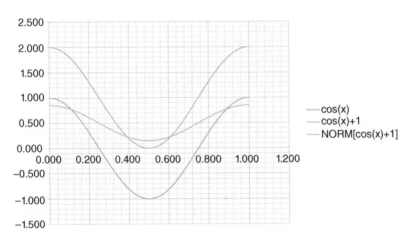

Figure 5.2 Cosine function – normalizing process.

After more than 3000 iterations, the conjugate gradient method finds an optimal solution associated with the values shown in Table 5.2 of weights, bias and mean square error.

By the conjugate gradient method, selectable in the *Solver* application, the error function is updated as the weights and biases change. The search for the minimum value of the error function can be interrupted by the user if he considers that the

Table 5.1 Cosine function.

NEURONS IN INPUT LAYER	1
HIDDEN LAYER	1
NEURONS IN HIDDEN LAYER	4
NUMBER OF OUTPUT NEURONS	1
ACTIVATION FUNCTION	SIGMOID
ITERATIVE METHOD	CONJUGATE GRADIENT METHOD
COST FUNCTION	MSE
NUMBER OF SAMPLES IN TRAINING SET	60
NUMBER OF SAMPLES IN VALIDATION SET	12
NEURON TEMPERATURE	1,00

Summary table of the neural network main characteristics.

Figure 5.3 Associated NN for cosine function model.

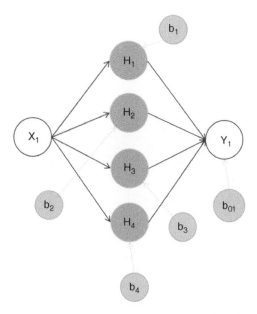

speed of the calculation of the minimum decreases step by step. The user realizes that the method has found the neighborhood of the function minimum. Alternatively, before the calculation the user can set an exceedingly high search time or an iteration number of the order of 10^5.

In such example, the error (MSE value in the Table 5.2), calculated with the training set, remains unchanged when the calculation is done for the validation set.

Table 5.2 Cosine function.

W_{11}	11.43	b_1	−1.52
W_{21}	−11.27	b_2	4.34
W_{31}	6.61	b_3	6.30
W_{41}	3.69	b_4	−1.47
W_{O1}	−5.84	b_{O1}	−15.83
W_{O2}	7.25		
W_{O3}	8.34		
W_{O4}	16.84	**MSE**	1.31E-0.5

Weight, Bias and MSE estimates.

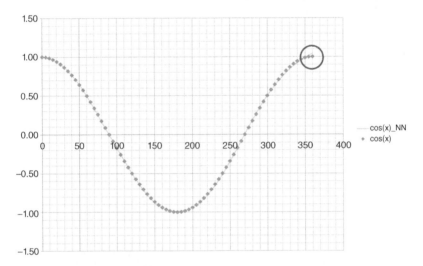

Figure 5.4 Results – comparison between real and predicted function.

Figure 5.4 shows how the function obtained from the network (yellow color) well represents the cosine function (blue points), apart from the last value near 360°. This aspect can be easily explained by the fact that the prediction is not dependent on other variables, (as will be seen in example 4), and above all no requirement of cyclicality has been sourced to the model for it to learn that the function is cyclic. Input neurons only defined the values of increasing degrees up to 360 that for values greater than 360 the network will find a solution along a horizontal asymptote.

The resulting function from the Neural Network is presented below:

$$y = \frac{1}{0.7} \frac{1}{1 + e^{-\left(-5.8 \cdot \frac{1}{1+e^{-(11.4 \cdot x - 1.5)}} + 7.2 \cdot \frac{1}{1+e^{-(-11.3 \cdot x - 4.3)}} + 8.3 \cdot \frac{1}{1+e^{-(6.6 \cdot x - 6.3)}} + 16.8 \cdot \frac{1}{1+e^{-(3.7 \cdot x - 1.5)}}\right)} + 15.9} - 0.15$$

The verification process, as per SE regulations, can:

- establish any corrective actions when a non-conformity is met,
- confirm that all system elements and the system of interest accomplish the intended functions, in accordance with the assigned requirements.

In the following example a verification activity is determined by the criticality of the considered system (cosine function software). Some mathematical artifacts will be used to helpfully define the cosine function. In fact, it is important to establish all of those verification criteria to create a valid procedure to confirm that the system works properly.

The outputs of the verification process are all the documentations of the verification results, a record of any recommended corrective action, feedback of corrective actions taken, and evidence that the system element or system meets the requirements, or not. For simplicity's sake, let's avoid documenting the verification results and put the data into a **Traceability Matrix**. A Traceability Matrix can easily be created by going indicate all the criteria used to define a cosine function through the NN usage.

For the cosine-system just designed the verification is as follows: the model is in compliance with all imposed requirements. An inconsistent requirement will lead to an incorrect verification of the model, therefore it is usual to consider those requirements which are formalized in the testing activities.

The validation process performs a comparative assessment as a means of evidence determining whether the stakeholder requirements and defined effectiveness measures have been correctly linked to the technical design specifications.

After the system of interest has been verified, it is submitted to the validation criteria. The previously established requirements tracking matrix is maintained.

The main outputs of the Validation Process are

- Results of the validation activity.
- Design feedback/corrective action.
- Approved system baseline.

The validation procedures demonstrate that the system is fit for purpose and meets stakeholder requirements

For the cosine-system just designed the validation is as follows: success-ful design verification. The validation of the model is satisfied when the estimated results are to be the same as that expected and thus conform to the user's expectations.

5.2 Example 2: Corrosion on a Metal Structure

A key aspect of the system life cycle is the maintenance process. A system, dur-ing its life-cycle, needs to be continuously monitored in its operating environment (See Figure 5.5) in order to identify and correct the failures by preventive actions which are able to restore the full system capability. Fulfilling the system's need for maintenance by establishing it early on as a non-functional requirement will allow us to save significant time and cost later on during the system operation, some of it can be achieved by accurate wholistic modeling of the system (see Model Based Systems Engineering – MBSE).

The purpose of this example is to show the application of NNs in a SE main-tenance process (See Figure 5.6). The maintenance of a metal structure can be introduced in the system life-cycle through the development of supportability cri-teria, hence the maintenance process corresponds to a phase of the life cycle called the supportability phase.

The metal corrosion is defined in the system specifications to ensure that it can be assessed early in the design phase, hence, based on the above, the **Inputs** must:

- support project requirements;
- improve the design of operational and future systems;
- use historical data and performance statistics to maintain high levels of reliabil-ity and availability.

Figure 5.5 An efficient design, but how many maintenance ships do we need to send to inspect? Source: Pixabay.

Figure 5.6 ANN in the SE maintenance process.

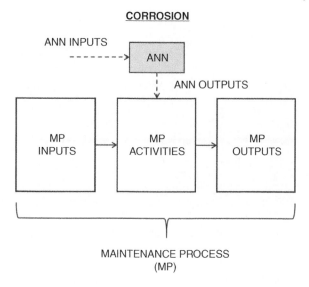

The maintenance process **Outputs**, on the other hand, must define some maintenance strategy and take into account the technical availability of the system such as:

- replacement of system elements,
- logistical support,
- training of maintenance personnel,
- fault reporting,
- actions to be taken in case of reported failures.

Once the results of the algorithm, built up by the NN, were analyzed, it is possible to prevent some of the **Activities** of the maintenance process. Certainly, we can define the following **Activities** starting from the NN outcomes:

- Establish a maintenance strategy.
- Define maintenance constraints on system requirements.
- Monitor spare parts replenishment levels.
- Manage the availability of trained maintenance personnel.
- Implement problem reporting and resolution procedures, including scheduled replacement of system elements prior to failure (preventive maintenance).
- Maintain failure history to outline possible maintenance solutions in other projects using similar systems.

In this example the NN provides context for identifying and treating two types of corrosion which can be found while inspecting a metal structure (See Figure 5.7 as a typical metal structure). The application of a neural network can be useful

Figure 5.7 A rusty iron bridge image by Jerzy Górecki from Pixabay.

for defining the above mentioned activities when the parameters of our corrosion model have been consistently set up. Our investigation should start from the phenomenon of corrosion in a metal structure, to create a reliable model (it is advisable to take notes and list the characteristics of the phenomenon to be analyzed).

Most metals, such as aluminum, are not used in pure form but they are combined with other metals to form stronger and more ductile alloys. Microscopically, metal alloys are made up of small crystalline regions, called grains, to form the metal part that is visible to our eyes. Corrosion can surface itself both on the exterior of these regions (surface corrosion), and/or, within the material, through the grains (inter-granular corrosion).

Corrosion can be accelerated at high temperatures. In this phase chemical reactions result in a concentration of water vapor in the air. The surrounding atmosphere and water are the most common corrosive agents (For example, sulfuric acid, halogen acids, nitrous oxide compounds and organic acids present in the waste of humans and animals), the presence of oxygen or water in the atmosphere, in the form of humidity, increases the corrosion process of metals in contact. Mixed with water we can also find quantities of mineral conductive dissolved inside. Conductivity has a direct effect on corrosion: the formation on the surface of an electrolyte solution mixed with acid gases, dirt and engine exhaust gases favor the electrical conductivity thus, increasing the corrosion rate. The metal around this area corrodes more quickly until it loses its strength. Finally, the surrounding atmosphere can also contain other corrosive and contaminating gases, in particular industrial products and marine sales, which can trigger chemical reactions.

The **surface corrosion**[17] is the most common type of corrosion, and it is the direct result of a chemical reaction occurring on the metal surface. As we mentioned earlier, if this corrosion is not removed quickly, an oxidation phenomenon can occur causing the origin of tiny cavities. Oxidation is the loss of electrons for metal molecules during a chemical reaction when oxygen interacts with the metal structure. These small superficial cavities can widen up to penetrate through the metal and cause a complete detachment of all the grains of initially damaged area, i.e., a fracture. According to the scientific community, conditions of relative air humidity between 78% and 90% can lead to **inter-granular corrosion**[18]. Additionally, the grains of the metal would react to each other and in some materials where the grain sizes are generally small, the microscopic interaction is accentuated causing a rapid detachment.

In this exercise the contribution of tension to the corrosive phenomenon and the contribution of fretting[49] due to vibrations are neglected, but we must not forget that metal structures are subject to external loads that are transmitted internally, that is, stresses. This stress can also occur during production or maintenance phase.

Corrosion increases in a harmful chemical environment if, in addition to the internal stresses of the metal, we also consider a state of constant or cyclic stress (fatigue i.e. passage of cars on a bridge or wind loading). By design, metals can withstand the cycle of stresses as much as the equivalent stress which is less than the resistance limit of the metal. However, when the part or structure is subjected to cyclic loading (fatigue[11]) and simultaneously exposed to a corrosive environment, corrosion damage can occur even at low stress levels. In this chemical–mechanical state, cracks can grow (fatigue cracks) and can propagate through the corroded part of metal. A fracture of a metal part due to fatigue corrosion generally occurs at a stress level well below the fatigue limit of a non-corroded part, even if the amount of corrosion present is relatively small.

Finally, in such example, no primer layer[35] on the metal surface is applied or any surface treatments such as anodic oxidation, which are intended to protect the metal from corrosion. The hypothesis we will make is of a generic metal structure in service for more than 20 years having occasional damage to the external protection.

Neural Network Input:

- Relative Humidity [11% – 90%]
- Temperature [10 °C – 34 °C]
- Zone to be inspected for corrosion: Zone 1 (0), Zone 2 (0.5), and Zone 3 (1)
- Hours in Service: [24.846–50.156]

while the outputs are related to the type of corrosion found:

- Surface corrosion = 0
- Inter-granular corrosion = 1

We will try to establish an inspection program depending on the damage type found in a certain time interval expressed in hours.

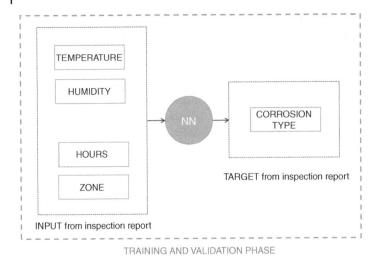

TRAINING AND VALIDATION PHASE

Figure 5.8 Scheme of the training and validation phase in a maintenance process activity.

Having defined the area to be inspected (primary structure and material) for the same structure configuration, we can set up an inspection program based on the experience acquired and developed by our neural network.

Furthermore, the program can be further customized by defining a range of temperature and relative humidity levels that metals would frequently operate in (See Figure 5.8 as a schematic approach).

- Input dataset is collected by inspecting a large amount of metal structures and will include the following information: type of corrosion.
- Inspection area.
- Total number of hours (connected to the time of finding).

The total number of hours is a temporal reference only, which will be useful for understanding how long it takes to generate surface corrosion rather than inter-granular corrosion in a certain climatic state.

Table 5.3 shows the main characteristics of the network built to define our model which is represented in the Figure 5.9.

Following more than 4000 iterations, the conjugate gradient method finds an optimal solution converted to values shown in the Table 5.4 of weights, bias and mean square error. Furthermore, the error (MSE value in the Table 5.4), calculated with the training set, remains unchanged when the calculation is done starting from validation set.

Our model can derive the type of corrosion on the three areas of the structure by varying the hours, setting two pairs of temperature and relative humidity values.

Table 5.3 Corrosion classification.

NEURONS IN INPUT LAYER	4
HIDDEN LAYER	1
NEURONS IN HIDDEN LAYER	9
NUMBER OF OUTPUT NEURONS	1
ACTIVATION FUNCTION	SIGMOID
ITERATIVE METHOD	CONJUGATE GRADIENT
COST FUNCTION	MSE
NUMBER OF SAMPLES IN TRAINING SET	70
NUMBER OF SAMPLES IN VALIDATION SET	14
NEURONS IN INPUT LAYER	1,00

Summary table of the neural network main characteristics.

Figure 5.9 Associated NN for corrosion classification.

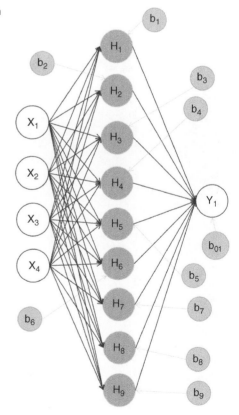

Table 5.4 Corrosion classification.

W_{11}	22.61	b_1	−14.17
W_{12}	−0.94	b_2	31.29
W_{13}	28.47	b_3	27.27
W_{14}	−25.64	b_4	−36.23
W_{21}	−2.80	b_5	−3.20
W_{22}	−9.14	b_6	−36.78
W_{23}	−56.74	b_7	4.54
W_{24}	3.49	b_8	4.50
W_{31}	2.74	b_9	5.68
W_{32}	−28.81	b_{O1}	−30.59
W_{33}	10.28		
W_{34}	−33.00		
W_{41}	18.77		
W_{42}	30.00		
W_{43}	11.76		
W_{44}	16.0		
W_{51}	2.30		
W_{52}	21.00		
W_{53}	32.00		
W_{54}	−16.66		
W_{61}	5.97		
W_{62}	32.02		
W_{63}	22.00		
W_{64}	14.00		
W_{71}	3.38		
W_{72}	5.12		
W_{73}	6.08		
W_{74}	0.75		
W_{81}	−9.42		
W_{82}	1.82		
W_{83}	12.77		
W_{84}	−14.93		
W_{91}	−0.02		
W_{92}	1.52		
W_{93}	−0.51		
W_{94}	−0.27		
W_{O1}	−37.30		
W_{O2}	−54.18		
W_{O3}	97.44		
W_{O4}	60.42		
W_{O5}	16.8430.81		
W_{O6}	45.54		
W_{O7}	−18.36		
W_{O8}	−17.21		
W_{O9}	−17.39	*MSE*	2.06E-0.7

Weight, Bias and MSE estimates.

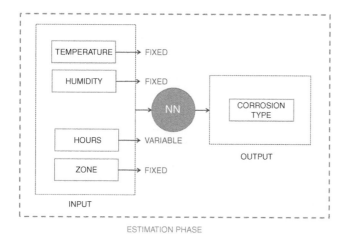

ESTIMATION PHASE

Figure 5.10 Scheme of the estimation phase in a maintenance process activity.

We can define two distinct cases starting from the average values of temperature and relative humidity referred to a certain geographical area where that type of metal structure is located (See Figure 5.10 as a schematic approach):

- Temp. = 25 °C, Humidity = 60%
- Temp. = 15 °C, Humidity = 25%

The summary of findings for the first case is:

- Zone 1 (See Figure 5.11)
- Surface corrosion inspected between 25 000 and 35 390 hours.
- Inter-granular corrosion inspected between 35 890 and 50 000 hours.

In the caption on the right, T indicates the temperature in degrees Celsius and Ur indicates the relative humidity.

- Zone 2 (See Figure 5.12)
- Surface Corrosion inspected between 25 000 and 31 890 hours.
- Inter-granular corrosion inspected between 32 390 and 50 000 hours.

In the caption on the right, T indicates the temperature in degrees Celsius and Ur indicates the relative humidity.

- Zone 3 (See Figure 5.13)
- Surface Corrosion inspected between 25 000 and 28 890 hours.
- Inter-granular corrosion inspected between 30 390 and 50 000 hours.

In the caption on the right, T indicates the temperature in degrees Celsius and Ur indicates the relative humidity.

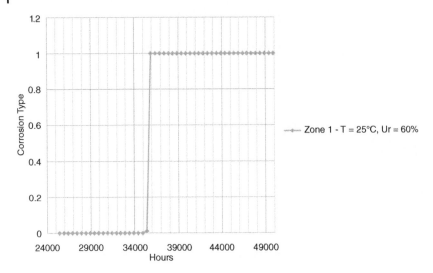

Figure 5.11 Zone 1 – From surface to inter-granular corrosion.

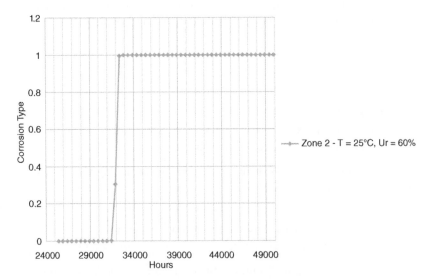

Figure 5.12 Zone 2 – From surface to inter-granular corrosion.

The result for the second case is:

- Zone 1 (See Figure 5.14)
- Surface Corrosion inspected between 25 000 and 38 390 hours.
- Inter-granular corrosion inspected between 38 890 and 50 000 hours.

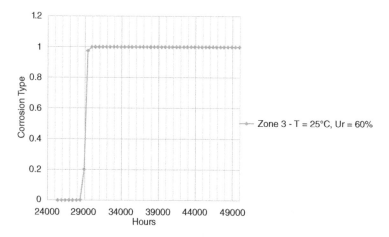

Figure 5.13   Zone 3 – From surface to inter-granular corrosion.

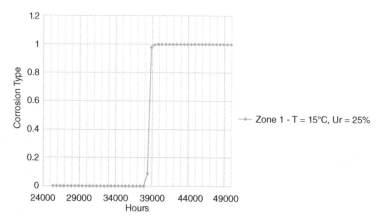

Figure 5.14   Zone 1 – From surface to inter-granular corrosion.

In the caption on the right, T indicates the temperature in degrees Celsius and Ur indicates the relative humidity.

- Zone 2 (See Figure 5.15)
- Surface Corrosion inspected between 25 000 and 37 390 hours.
- Inter-granular corrosion inspected between 37 890 and 50 000 hours.

In the caption on the right, T indicates the temperature in degrees Celsius and Ur indicates the relative humidity.

- Zone 3 (See Figure 5.16)
- Surface Corrosion inspected between 25 000 and 42 390 hours.
- Inter-granular corrosion inspected between 42 890 and 50 000 hours.

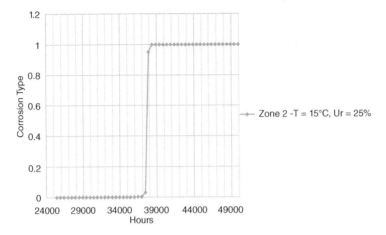

Figure 5.15 Zone 2 – From surface to inter-granular corrosion.

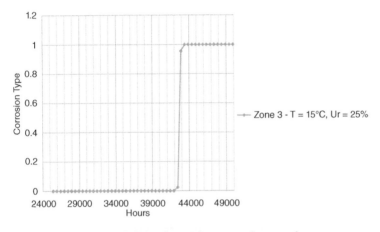

Figure 5.16 Zone 3 – From surface to inter-granular corrosion.

In the caption on the right, T indicates the temperature in degrees Celsius and Ur indicates the relative humidity.

The graphs are the result of a simplified scenario aimed at indicating the time of corrosion on a certain area of the structure, and certainly they do not indicate the inter-granular corrosion that always appears after the surface corrosion. As mentioned, we can confirm that the network has collected real data and imposed a mathematical model to classify them and provide how many hours one or the other corrosion occurs. Furthermore, we remind you that we are not inspecting a detailed area of the metal structure, but we are stating that within a given time frame corrosion occurs on a macro area. An extension could be to specify sub zones to define a more precise time interval and inspection area.

5.3 Example 3: Defining Roles of Athletes

In Chapter 3 we already introduced a sports business scenario, this example as well as the next will make use of that as playing field for our experimentation on neural networks (NN).

The selection and the choice of the best resources and consequently their inclusion in a project is addressed by the Resource Management Process (RMP). This RMP is necessary for the efficient and effective allocation of resources within an organization. Such process allows for the evaluation of the skills and performance of Human Resources, with the goal of optimizing the use of each resource while minimizing costs, since it is unlikely to achieve full utilization of each resource while minimizing expenses. RMP is heavily influenced by anticipated needs over time. Through the use of NNs and related mathematical models, it is possible to accurately identify the role of each resource and give insight into their performances.

The NN, inserted in the process of resource management, allows us to collect in an active portfolio all the data needed for the project. The resources can be chosen based on technical and tactical requirements, with the goal of providing the team with what is needed to keep the project on target. A concern is to prevent the team's athletes from becoming overcommitted; their utilization can be optimized when their skills are critical to a specific technical/tactical requirement. Skills inventories are important documentation that can be validated by the Project Manager (PM).

In this example we will present using a neural network model to work on a typical classification task. Athletes of widely common sports (e.g. soccer, American football, basketball) will be taken as reference, and we will aim for the model to be able to understand whether the athletes will have an offensive or defensive role based on their characteristics.

The input data as "Athlete characteristics" are:

- Decisive interceptions in both phases, offensive and defensive (See Figure 5.17).
- Decisive passes in both phases, offensive and defensive phases (See Figure 5.18).
- Distance covered (See Figure 5.19).

In this exercise we will normalize the input values while we encode the output values as follows:

- Defensive = 0
- Offensive = 1

Accurate definition and quality processing of input and output data are imperative in building an effective model, as the neural network model will infer based on the provided information.

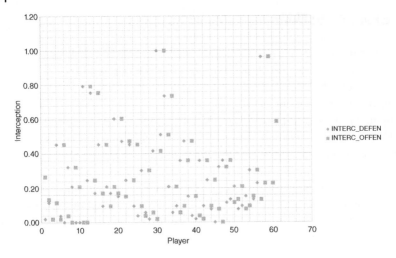

Figure 5.17 Scatter plot – interceptions.

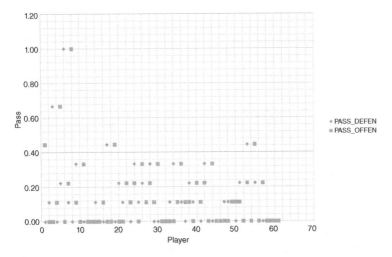

Figure 5.18 Scatter plot – passes.

The input values have been collected exactly (no probabilities).

The number of athletes is shown on the x-axis, while the average performances recorded during matches are shown on the y-axis. It is evident that a part of the total number of athletes becomes our data for training set while the remaining becomes the data for validation set. In addition, the role defined as offense and defense is usually obtained from the roster table.

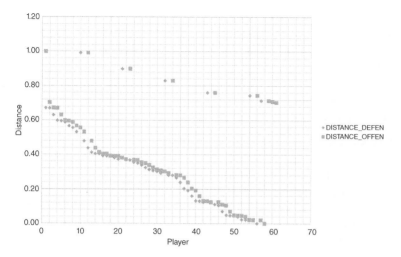

Figure 5.19 Scatter plot – distance covered.

From the scatter plots we can immediately understand that the only characteristic of intercepts and/or decisive passes would not help us to correctly write a classification algorithm due to low variance and presence of outliers. Therefore, the distances covered in the field (which can be evaluated, for example, with a thermographic approach) reinforce the classification process with useful insights.

As we have shown before, the input data have been made proportional to the transfer function since the solution will not be far from the optimal weight. The difference between initial and final weights/biases will be smallest.

Table 5.5 shows the main characteristics of the network created to define the classification. In the Figure 5.20 the NN model is represented.

Furthermore, the error calculated starting from the training set is of the order of 10^{-5} while the calculation with the validation set increases up to the order of 10^{-2}.

Following more than 4000 iterations, the conjugate gradient method finds an optimal solution as shown in the Table 5.6 of weights, bias and mean square error.

As shown in the Figure 5.21, not all values correspond to the target set, 16 of 62 athletes display a small variation, which could depend on various factors that would require further evaluation. The instructions given during a match and other tactical aspects could affect the role of an athlete. For example, a typical offensive athlete may show to have, effectively, a defensive role during the matches analyzed.

Table 5.5 Defining roles of athletes.

NEURONS IN INPUT LAYER	3
HIDDEN LAYER	1
NEURONS IN HIDDEN LAYER	6
NUMBER OF OUTPUT NEURONS	1
ACTIVATION FUNCTION	SIGMOID
ITERATIVE METHOD	CONJUGATE GRADIENT
COST FUNCTION	MSE
NUMBER OF SAMPLES IN TRAINING SET	50
NUMBER OF SAMPLES IN VALIDATION SET	11
NEURONS IN INPUT LAYER	1,1

Summary table of the neural network main characteristics.

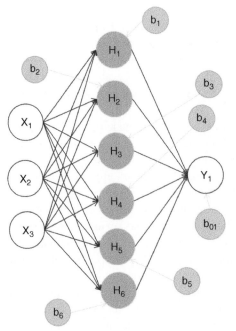

Figure 5.20 NN associated with roles classification.

Table 5.6 Defining roles of athletes.

W_{11}	*−30.68*	*b1*	*21.01*
W_{12}	81.83	b_2	3.86
W_{13}	−68.88	b_3	−13.23
W_{21}	−6.81	b_4	−20.27
W_{22}	−6.95	b_5	29.21
W_{23}	1.38	b_6	−2.15
W_{31}	114.65	b_{O1}	45.23
W_{32}	−67.25		
W_{33}	−35.01		
W_{41}	42.59		
W_{42}	116.03		
W_{43}	36.52		
W_{51}	−52.04		
W_{52}	77.87		
W_{53}	−30.22		
W_{61}	11.45		
W_{62}	3.20		
W_{63}	4.37		
W_{O1}	−66.78		
W_{O2}	−63.27		
W_{O3}	73.32		
W_{O4}	60.68		
W_{O5}	111.23		
W_{O6}	−146.22	*MSE*	1.66E-0.2

Weight, Bias and MSE estimates.

This exercise shows that we can use a neural network to identify any variations from the basic characteristics of an athlete. Some athletes may also play a different role than their main characteristics, and, from a strategic point of view, they can cover more phases of the game depending on the needs encountered during a competition.

As interesting annotation we have entirely omitted any of the psychological aspect from our analysis, reserving this topic to a future research. Things such as goal setting, personal awareness, routine, concentration, meditation, confidence, control, and even overall flow mindset can determine whether an athlete will perform poorly or exceptionally in their chosen field.

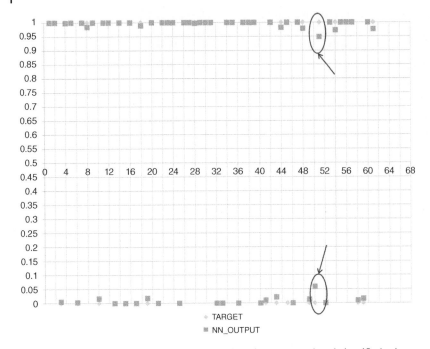

Figure 5.21 Scatter plot – results – comparison between real and classified roles.

5.4 Example 4: Athlete's Performance

Once again the system engineering processes, Resource Management Process (RMP), mentioned in the previous example, in particular, comes to our aid in managing a sport club. The technical/tactical demands may suggest some changes in the team composition, so the sport club can avail itself of expert evaluations to provide possible actions for performance improvement and reward. A typical example of sport club resources is shown in the Figure 5.22.

NOTE: Balancing team expertise against other elements of the sports organization (e.g. IT tools, medical tools, gyms, training field), and maintaining a balance between the project budget and the cost of resources, is a complicated duty entrusted to a qualified personnel such as a Project Manager (PM). A PM must manage a lot of variables and must keep abreast of policies and procedures, ready to modify them if necessary.

An athlete's performance prediction, incorporated into the RMP, provides patterns for the PM to be used in the following activities:

- Manage resource availability to ensure that club goals are met.
- Manage resource shortages through recommendations and issue solutions.

Figure 5.22 Using neural networks big data to predict match outcomes? We can try. Source: Pixabay.

- Manage requests for new talent.
- Manage the recruitment and training of an athlete based on experience levels, skills, motivation, leaderships, etc.
- Develop a training or performance improvement policy on a regular cycle.
- Create an information infrastructure for athlete management. Support systems to maintain, track, allocate and improve athlete performance targeted to club needs.
- Maintain an athlete's career development program that is not diverted by project needs.

In this exercise a predictive approach for an athlete's performance is shown. The neural network will process a high volume of data that will give a predictive result trended to a desired target. As previously mentioned, it is important that the available data are translated to its simplest and appropriate forms to guarantee a clear convergence.

The input data are:

- Number of played matches.
- Trend over time of the athlete's psycho-physical and tactical state.

While in output the score is analyzed over 172 matches (See Figure 5.23 for cumulative score).

For simplicity of this exercise, we have incorporated in a single value as the psycho-physical and tactical state, according to an oscillatory trend. (It is unlikely that an athlete will be able to maintain a constant of level of good condition). Such data could be body mass, stamina, speed, age, position on the field and other mental aspects that characterize the mood, self-confidence, desire to improve their performance.

In this case, the theory of Fuzzy Logic can be useful to discretize qualitative data into quantitative ones. However, you would require support from specialists of the field for a reliable model. Players' performances are summarized in the

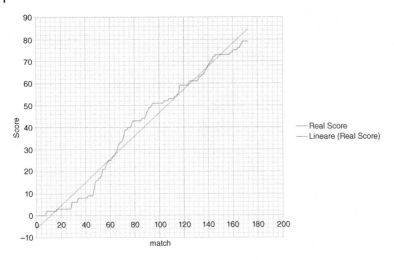

Figure 5.23 Target function – real scores (cumulative).

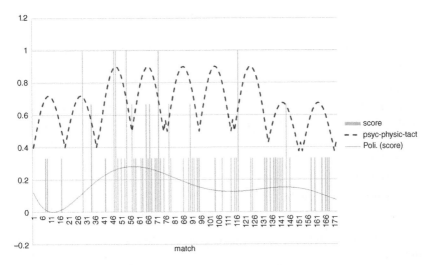

Figure 5.24 Normalized scores – normalized psyc-physic-tact status.

Figure 5.24 below (the number of matches on the x-axis and the normalized value of score on the y-axis) and the psycho-physical and tactical condition recorded in each match.

Normalization enables direct juxtaposition of multiple features for comparison even for those which may be seeming irrelevant.

Table 5.7 the main characteristics of neural network are reported. In the Figure 5.25 a NN model is represented.

Table 5.7 Athlete's performance.

NEURONS IN INPUT LAYER	2
HIDDEN LAYER	1
NEURONS IN HIDDEN LAYER	8
NUMBER OF OUTPUT NEURONS	1
ACTIVATION FUNCTION	TANH
ITERATIVE METHOD	CONJUGATE GRADIENT METHOD
COST FUNCTION	MSE
NUMBER OF SAMPLES IN TRAINING SET	142
NUMBER OF SAMPLES IN VALIDATION SET	30
NEURONS IN INPUT LAYER	1,6

Summary table of the neural network architecture.

Figure 5.25 NN associated with athlete's performance prediction.

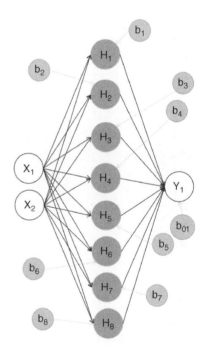

Once the curve is known, the model can be applied to other athletes, assuming that the training and evaluation techniques of the psycho-physical and tactical aspects are equivalent to those applied to the athlete under consideration (the data used to train the model). In fact, having a reference player, in the same working scenario, will lead to the creation of more reliable mathematical model.

In this regard, we do not exclude the possibility that this approach can also be applied to other scenarios, such as employees of a company or an aircraft engine.

After more than 10 000 iterations, the conjugate gradient method finds an optimal solution as shown in the Table 5.8 of weights, bias and mean square error.

Updating the weights and biases of the neural network through the training data generates a random variation of the gradient descent. In this example the training phase has been repeated in order to increase the chance of finding or approaching the global minimum of the cost function.

This Neural Network is highly nonlinear and the associated cost function have many local minima, so the gradient variation has been checked for each epoch in order to ensure the reaching to the global minimum.

Table 5.8 Athlete's performance.

W_{11}	−32.72	b_1	30.09
W_{12}	2.44	b_2	−160.06
W_{21}	187.14	b_3	−158.61
W_{22}	104.11	b_4	0.83
W_{31}	180.11	b_5	−102.93
W_{32}	106.24	b_6	0.00
W_{41}	1.31	b_7	−304.36
W_{42}	0.87	b_8	−0.23
W_{51}	−0.06	b_{O1}	16.22
W_{52}	356.21		
W_{61}	4.82		
W_{62}	−0.22		
W_{71}	409.59		
W_{72}	97.26		
W_{81}	0.11		
W_{82}	0.038		
W_{O1}	−52.60		
W_{O2}	2.71		
W_{O3}	−2.78		
W_{O4}	18.17		
W_{O5}	0.01		
W_{O6}	17.09		
W_{O7}	0.11		
W_{O8}	−5.99	*MSE*	3.84E-0.2

Weight, Bias and MSE estimates.

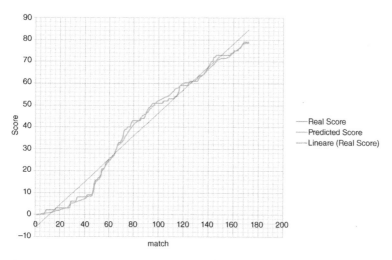

Figure 5.26 Results – comparison between real and predicted scores.

It is important to note that the function of the trend line in the real case coincides with that in the predicted case (See Figure 5.26). It is also evident that between the hundredth and one hundred twentieth match, the performances do not coincide: in the real case it is a constant (performance) phase while in the predicted case the curve is ascending. The explanation must be sought for any injuries or minutes played in each analyzed match, the player's psycho-physical and tactical status. We have specifically kept these results to highlight having a precise definition of the inputs. For example, entering the minutes played in each match rather than the number of matches, has an important impact on the predictive event we want to estimate. In conclusion, below charts show that real and expected trend share the same gradient (See Figures 5.27 and 5.28).

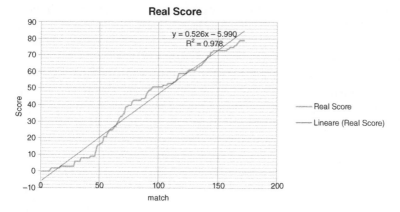

Figure 5.27 Linear function – real score.

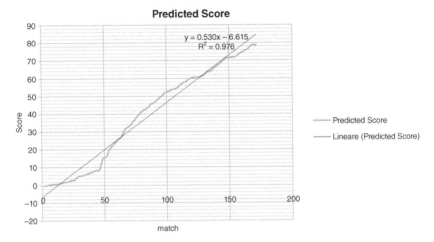

Figure 5.28 Linear function – predicted score.

5.5 Example 5: Team Performance

5.5.1 A Human-Defined-System

From the definition of systems in ISO/IEC 15288, systems, thought of as products or services, are human-made, and configured with one or more of the following parts: hardware, software, human processes, procedures and structures. A SE (Systems Engineering) discipline focuses on the ability of an organization to manage a human-defined system.

An organization manages a system through processes designed to establish and maintain a set of policies and procedures that support the organization's ability to acquire and deliver products and services. The integration of processes that support human resources (such as finance) are critical to achieving the organization's strategic goals. Paying attention to human factors means refining and continually improving the business environment and understanding the relationships and communications between parties.

Project managers are tasked with balancing resources between the less experienced and the more experienced, knowing that the project depends on teamwork and an optimally multidisciplinary team. Such teams are able to solve problems quickly through direct communication, which shortens the decision cycle and is more likely to result in better decisions.

In a multidisciplinary team, each member is an expert in his or her own discipline, but risks having his or her own perspective stand out above the others. For example, to achieve a sports team's goal, one member considers their specialty indispensable at the expense of another. The team could be doomed to mediocrity unless each member approaches the work with challenging ideas, focusing on the end result. Therefore, a project manager has the responsibility to represent all points of view, and establish the necessary relationships among all team members. First and foremost, project managers need to balance between experienced and novice resources to achieve team stability. Today's projects depend on teamwork and optimally multidisciplinary teams. The balanced team is able to solve problems quickly through direct communication among its members. Such intra-team communication shortens the decision-making cycle and is more likely to result in better decisions. However, studies have shown that team decisions are often more "risky," resulting in the potential for greater innovation.

In a multidisciplinary team, each member comes from a discipline with their own perspective. They are responsible for representing that point of view, while at the same time necessary relationships with other members. However, team results are doomed to mediocrity unless each member approaches the work with challenging ideas while focusing on the end result. System projects depend on the effective integration of multidisciplinary efforts.

It is recommended that the organization of a systems project should provide an opportunity for all disciplinary specialists to work together continuously face-to-face and, most importantly, to gain systems perspective and an understanding of the role their specific knowledge can provide in deriving a particular systems design.

Different phases of the life cycle require different tasks and different staff skills. This allows management to acquire and properly utilize the right combination of specialized and generalist skills. A project avoids the "bureaucratization" by streamlining the organization and integrating the various specialist backgrounds into common system-oriented teams with direct loyalty to the systems design effort.

Modern projects use the concepts of integrated product and process teams to establish a project organization.

5.5.2 Human Factors

According to the Human Factor discipline developed in the aviation sector, and based on ICAO Circular 227, the study of human factors[22] aims to find the best conditions for safety and efficiency when workers carry out their tasks in both physical and interpersonal sense. Each worker must be proficient in the work equipment and follow the procedures in a correct and disciplined manner. Accident investigators have identified the human element as a major source of errors and the main causes of most accidents. The main elements studied by the experts were:

- The nature of the error.
- The triggering causes.

- Psycho-physical limitations.
- Human relationships.
- Error measures.

The physical and psychological limits are certainly linked to the performance assessed during the work activities to establish the causes leading to the error, such as tiredness, lack of knowledge of the tasks to be performed or poor communication in the working group. In fact, the interaction between various members of the group, and therefore all aspects concerning human relationships, such as leadership, conflicts and criticism, are addressed through targeted exercises such that members are aware of their behavior when working as a group.

Once a root cause is known, decision-making process can be directed toward training and ergonomics. Training is used to detect subtle, latent dangers, while the study of ergonomics, i.e. the analysis of concepts linked to the use of mechanical or electronic systems, is necessary to improve their use.

In short, when we talk about the human factor, the following questions must be answered:

- why and how do humans make mistakes?
- how much does human error affect the final result?
- what is the relationship between humans and technology?

It is evident that this discipline can certainly be applied in fields other than aeronautics. We therefore believe that it is possible to study the influence of psychological aspects in the sport business, starting with an in-depth examination of the human factor, i.e. identifying mistakes of an athlete. Reducing these errors would allow improvement in an athlete's performance.

5.5.3 The Sports Team as System of Interest

In this example, a source of potential errors is classified starting from the numerical formalization of the error measurement and the behavior of the athlete during the match and training phases.

Input (array):

- Fouls.
- Yellow cards received.
- Red cards received.
- Off-sides.
- Handball.
- Conceded penalty to the opponent.

Output (array):

- Win.
- Loss.
- Draw.

In an initial analysis, it is possible to evaluate the incidence of the cumulative team errors toward the final score of a match. It is evident that the algorithm can be developed to derive the patterns affecting the following causes (not limited to):

- Lack of training.
- Lack of tactical knowledge.
- Lack of communications.
- Lack of stamina.

We believe that a further analysis, carried out with experts using the Delphi method, can customize the algorithm. Anyway, in the Table 5.9 the main characteristics of used NN are reported.

Note: For brevity many elements of this examples have been removed, but the whole Google Colab file can be requested from http://www.ai-shed.com

5.5.4 Impact of Human Error on Sports Team Performance

As discussed, human error is an undeniably important factor the impacts performance of sports activities. In this exercise, we would like to evaluate this by analyzing matches played by various teams across multiple football leagues.

5.5.4.1 Dataset

This dataset has been derived from Football Events dataset by Alin Secareanu in Kaggle.

Feature engineering and extraction have been applied in the original dataset to create input data. The interest is on quantitative data which can represent human error, the impact of accumulated error factor in the overall team performance hence in the final match result.

5.5.4.2 Problem Statement

We would like to understand whether human error in the match could make contribution in the outcome of a game.

Table 5.9 Team performance.

NEURONS IN INPUT LAYER	301
HIDDEN LAYER	2
NEURONS IN HIDDEN LAYER	10
NUMBER OF OUTPUT NEURONS	3
ACTIVATION FUNCTION	Several, see code
ITERATIVE METHOD	GRADIENT
COST FUNCTION	MSE
NUMBER OF SAMPLES IN TRAINING SET	See code
NUMBER OF SAMPLES IN VALIDATION SET	See code

Summary table of the neural network architecture.

Assumptions have been made that all errors regardless of the timing in the game will have equal impact on the target variable considering that no time variable is considered in this modeling. Hence, the area of improvement on the model will be using time series data of activities which would also take into account the order of activities occurred in a match.

5.5.4.3 Feature Engineering and Extraction

The first step us to extract and then to organize the dataset:

- create a new column to represent the target variable: match result
- use functions to generate new computed columns

The following is a list of preliminary operations (code omitted for brevity) on the dataset:

- reads match metadata *.csv* file;
- drops unnecessary columns;
- filter team match results only (which has the most number of match records, both in home and away games);
- creates a new column for full time result in semantic labels (home, away, draw);
- creates a new column called "home_away" to flag if a game was a home game or an away game;
- creates a new target column which will denote win/loss or draw;
- swap the values in home goal and away goal: count to have the number of goals scored by team;
- update the column names with more semantically accurate ones.

5.5.4.4 Creation of Computed Columns

The computed columns will display total count of each human error inflicted events made by their player. Considering the analysis of one team, we will aggregate the counts from events dataset and record in the match dataset (See Table 5.10).

Table 5.10 Extract of the data with additional columns created.

goal	opponent_ goals	match_ result	home_ away	result	foul	yellow_ cards	red_ card	Off-sides	Hand ball	allowed_ penalty
46 4	0	home	home	win	6	0	0	0	0	0
73 3	2	away	away	win	7	1	0	0	1	0
107 5	1	away	away	win	3	0	0	0	0	0
138 3	0	home	home	win	2	0	0	0	0	0
199 2	2	draw	away	draw	8	1	0	1	0	0

The columns include:

- Total number of fouls.
- Total number of yellow cards received.
- Total number of red cards received.
- Total number of off-sides.
- Total number of handball.
- Total number of times conceded penalty to the opponent.

5.5.4.5 Explorative Data Analysis (EDA)

From the explorative analysis of data considered in the first stage, non-contributive columns can be replaced by other columns having more meaningful values. Collecting too many measures, without knowing how they can be used, wastes time and resources, and even worse, the truly useful values may get confused among the large amount of accumulated data (See Tables 5.11 and 5.12). Therefore, it is desirable that amount of accumulated data be limited to truly critical issues. If you have downloaded these notebooks to your drive, such can be mounted to this colab notebook in order to access the input data file. "drive.mount" command will open up a separate window where you should authenticate with your account.

Let us see the distribution of output labels (home, away and draw), these are stored in the column "ftr." The distribution suggests that the dataset is imbalanced where there is relatively more number of matches where home team has won the match compared to away team winning with getting a draw being the lowest (See Figure 5.29).

```
ftr_count = df.ftr.value_counts()
ftr_count.plot(kind='bar', title= 'Count (home away games)')
```

Preprocess dataset

Split the dataframe to input and target variables. This will be randomly split between train and test sets with 15% of total dataset to be kept for testing purpose post training.

Table 5.11 Extract of the data (part 1) for EDA process.

	ht	at	goals	oppon_ goal	Foul_ ht	Foul_ at	yellow_ cards_ht	yellow_ cards_ht	Red_ card_ht	Red_ card_at
0	Team 1	Team 2	3	1	12	17	1	3	0	0
1	Team 3	Team 4	2	2	15	25	2	4	0	0
2	Team 5	Team 6	2	0	12	25	1	2	0	0
3	Team 7	Team 8	0	1	20	8	1	3	0	0
4	Team 9	Team 10	1	0	10	10	2	1	0	0

Table 5.12 Extract of the data (part 2) for EDA process.

	ht	at	Off-sides_ht	Off-sides_at	Bad_shot_ht	Bad_shot_at	Missed_pass_ht	Missed_pass_at	ftr
0	Team 1	Team 2	1	3	11	6	0.44	0.25	Home
1	Team 3	Team 4	1	0	4	5	0.27	0.22	Draw
2	Team 5	Team 6	6	6	14	6	0.21	0.22	Home
3	Team 7	Team 8	5	6	9	5	0.50	0.22	Away
4	Team 9	Team 10	2	1	8	6	0.54	0.25	Home

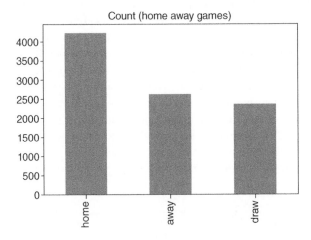

Figure 5.29 Count total amount of match results.

Data Preprocessing Steps

Update the data types to make them consistent across the input features.

```
# Update column types, cat to category and numerics to
 floats
df['ht'] = df['ht'].astype('category')
df['at'] = df['at'].astype('category')
df['ftr'] = df['ftr'].astype('category')

# Update all numerical columns to have type float
int_cols = df.select_dtypes(exclude=['category']).columns
df[int_cols] = df[int_cols].apply(pd.to_numeric,
 downcast='float', errors='coerce')
```

Split the data to input features (X) and target variable (Y).

```
# Split to X and Y sets
Y = df['ftr']
X = df.loc[:, df.columns != 'ftr']
```

We will apply encoding for target variable Y which will be in two steps.

- Apply Label Encoder to **encode text class names to numerical values**
- Apply OneHot Encoder to translate single numerical values to have n-dim **binary representation** where n is the number of output classes.

Split the input data into train and test set, where the test set will have 15% of the total dataset. This will be kept aside until predictions can be made with a trained model.

```
from sklearn.model_selection import train_test_split
from sklearn.pipeline import Pipeline
# Split to train and test sets
X_train, X_test, y_train, y_test = train_test_split(X,Y_le,
test_size=0.15, random_state=0)
```

We will use a ColumnTransformer to apply scaling and encoding on numeric and categorical input features respectively. The transformer will be fitted to the training data only. Having a transformer is useful as it can also be reused to apply the same transformation on testing data.

```
# Apply Scaling for numeric columns
num_features = ['foul_ht', 'yellow_cards_ht', 'red_card_ht',
 'offsides_ht',
 'handball_ht', 'allowed_penalty_ht', 'bad_shot_ht',
 'missed_pass_ht',
                'foul_at', 'yellow_cards_at', 'red_card_at',
 'offsides_at',
             'handball_at', 'allowed_penalty_at',
 'bad_shot_at', 'missed_pass_at']
cat_features = ['ht', 'at']
# Apply one hot encoding for categorical features.

ct = ColumnTransformer([
        ('num', RobustScaler(), num_features),
        ('cat', OneHotEncoder(handle_unknown='ignore'),
         cat_features),
    ])

X_train = ct.fit_transform(X_train)
X_train = X_train.astype(float)
```

5.5.4.6 Extension - Sampling Method for an Imbalanced Dataset

As seen in the above Figure 5.29, we have an imbalanced dataset where one class holds substantially more records than the remaining two with the least only holding almost a third of the majority. In this case, sampling is one of the ways to mitigate the issue coming from the imbalanced dataset.

Considering the size of the dataset, we could either oversample the minority or undersample.

5.5.4.7 Building a Neural Network Model

We will build out a rather light multi-layer neural network model to train on. The Base model function will construct the model with a combination of Dense and Dropout layers to prevent overfitting. For more information of the functionality of the Keras library Dense and Dropout layer please consult https://keras.io/api/layers/core_layers/dense. In short Dense creates the network and dropout turn off selectively certain neurons in the hidden layer to improve efficiency.

```python
import tensorflow as tf
from tensorflow.keras.models import Sequential
from tensorflow.keras.layers import Dense, Dropout
from tensorflow.keras.wrappers.scikit_learn import
 KerasClassifier
from sklearn.metrics import confusion_matrix
from tensorflow.keras.metrics import Precision, Recall, AUC,
 CategoricalAccuracy
from tensorflow.keras.optimizers import Adam

n_features = X_tl.shape[1]



def baseline_model(loss='categorical_crossentropy',
 metrics=METRICS, output_bias=None):
    if output_bias is not None:
        output_bias = tf.keras.initializers.Constant
(output_bias)
    model = Sequential()
    model.add(Dense(301, activation='elu',input_dim=301))
    model.add(Dense(10, activation='elu'))
    model.add(Dropout(0.4))
    model.add(Dense(10, activation='sigmoid'))
    model.add(Dropout(0.4))
    model.add(Dense(3, activation='softmax'))
    # compile model
    model.compile(loss=loss, optimizer=Adam(lr=0.001),
                  metrics=[metrics])
    return model
```

Extension - Use of Learning rate decay and Class weights

As another mean of overfitting, Class weights are used as part of model training. This will help prevent the model being biased toward most dominant target class.

To support optimization, learning rate decay callback is used to reduce the learning rate as the epoch increases.

```
import sklearn
from sklearn.utils.class_weight import compute_class_weight
weight = compute_class_weight(class_weight='balanced',
 classes=np.unique(Y), y=Y)
weight = {i : weight[i] for i in range(3)}
esti1 = baseline_model()
callback_lr_decay = lr_decay_callback(0.001, 0.5)

X_tl_a = X_tl.toarray()
y_tl_a = y_train.toarray()
his1 = esti1.fit(X_tl_a, y_tl_a, epochs=100, batch_size=100,
 class_weight=weight, validation_split=0.15,
 callbacks=[callback_lr_decay], verbose=2)
```

5.5.4.8 Training Outcome and Model Evaluation

Assess the model training by understanding training results data from the history object, including variations of loss value and other metrics.

The learning curve (See Figure 5.30) relates the performance of the training phase to that of the validation phase over a number of epochs. The progression stabilizes on the plateau when the neural network has reached full capability. In this case, we see that the neural network doesn't improves its learning and validation performance. This aspect could indicate that the neural network has reached a limit of its capabilities (See Figure 5.31) or is no longer able to learn due to limited or redundant data input.

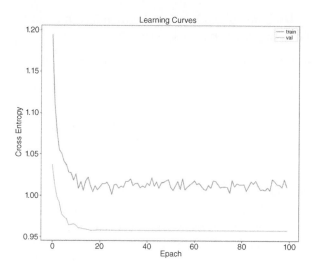

Figure 5.30 Learning curve: Cross Entropy variation.

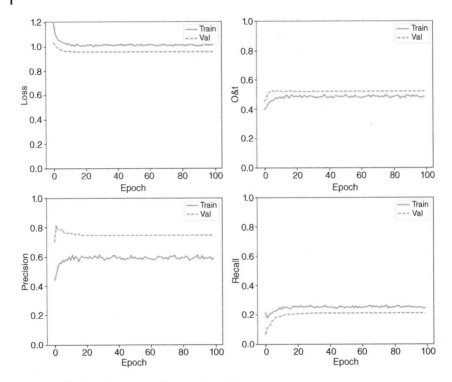

Figure 5.31   Learning curves: loss and precision.

The metrics generates above are also built in into Keras library and can be retrieved by adding the following code to run on the network.

```
# a set of metrics will be recorded for each epoch during training
METRICS = [
      Precision(name='precision'),
      Recall(name='recall'),
      AUC(name='auc'),
      AUC(name='prc', curve='PR'), # precision-recall curve
      CategoricalAccuracy(name='cat')
]
```

5.5.4.9   Evaluate Using Test Data
A confusion matrix describes the performance of the classification model. In other words, confusion matrix is a way to summarize classifier performance. Figure 5.32 shows a basic representation of a confusion matrix. Apply the column transformer on the input features of the testing set. Using the standardized input features, the model will predict the outcome for each match data. Use

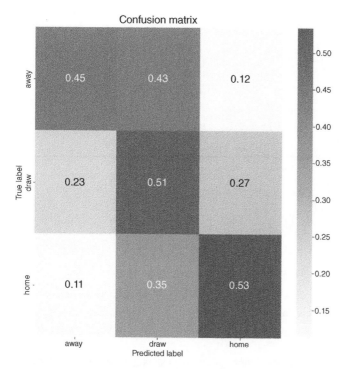

Figure 5.32 Confusion matrix: rows (true labels) and columns (predicted labels).

the function "get_confusion_matrix" to plot confusion matrix. Monitoring the confusion matrix means monitoring the training progress.

```
X_test_std = ct.transform(X_test)
y_pred = esti1.predict(X_test_std)
y_test = y_test.reshape(-1,1)
y_pred_maxed=np.argmax(y_pred, axis=1)
y_test_maxed=y_test

def get_confusion_matrix(y_true, y_pred, show,
 normalise=True):  # output of get_y_prediction
    dico = le_name_mapping
    cm = confusion_matrix(y_true, y_pred)
    orig_cm = cm
    if show:
        plt.figure(figsize=(12,12))
    if normalise:
        cm = cm.astype('float') / cm.sum(axis=1)[:,
np.newaxis]
        fmt = '.2f'
    else:
        fmt = 'd'
```

```
    # if 'cm_type' in kwargs:
    #     unlabelled = kwargs['cm_type']
    unlabelled = False
    classNames = [str(x) for x in list(dico.keys())]
    cm = pd.DataFrame(cm, columns=classNames,
index=classNames)
    if not unlabelled:
        plt.title("Confusion matrix")
        sns.heatmap(cm, annot=True, fmt=fmt, cmap='Blues')

        plt.ylabel('True label')
        plt.xlabel('Predicted label')
    else:
        sns.heatmap(cm, cmap="YlGnBu", xticklabels=False,
yticklabels=False)
    if show:
        plt.show()
    return orig_cm
```

```
get_confusion_matrix(y_test_maxed, y_pred_maxed, show=True)
```

This is a normalized view of confusion matrix where you can see the relative proportions of true labels data points predicted to each target class. So every row will sum up to 1. You can also see the underlying count of each predictions as un-normalized view.

We see that the current model is not one of the best as it performs just above 1/3 random chance in predicting the outcome of a match.

The classification report displays another set of useful metrics for evaluating a model trained with an imbalanced dataset. It shows precision, recall and f1 score. We would focus on the f1 score than the usual accuracy as it takes into account each individual class performance. The figures align with what we see from the confusion matrix that it performs slightly above pure random guessing with an average of 50% accuracy across all the target classes.

There indeed exist lots of areas of improvement from more data points and better represented data with time bounds, model architecture search, and hyper-parameter tuning.

5.6 Example 6: Trend Prediction

Moving away from our sports context we tackle another problem often solved with Neural Networks, the prediction. In this case it will be applied to the global crisis raging while this book was being prepared.

 Before explaining the example, it is important to note that the study of pandemics goes back many centuries. In general, a pandemic is when a disease spreads rapidly and affects people or animals far from the place where the first case of contagion is found. Throughout history the best-known pandemics are the plague and the "Spanish" flu which caused over 50 million deaths worldwide. Only with the advent of AIDS and later SARS (Severe Acute Respiratory Syndrome) it was possible to develop models for the control and prediction of infectious diseases.

We have repeatedly stated that the study of an event is based on the concept of a multidisciplinary approach, and similarly for epidemiology, we must add several disciplines: medicine, microbiology and mathematics. The first epidemiological model was developed by Bernoulli, a well-known Swiss mathematician, who gained notoriety for his most famous theorem. Based on the concept of likelihood, he established a hypothesis of constant proportionality, according to which the risk of contracting a disease is a function of the age of the uninfected person. Conversely, another constant was calculated as a function of the number of infected persons, establishing an estimate of how many will die once they contract the disease. Only centuries after Bernoulli's work, the British physicians J. Snow, and Nobel prize winner, Sir R. Ross, laid the foundations of modern epidemiology, i.e. introducing the concept of data acquisition and parameters processing for the first time.

Snow was able to establish relationships between events and formulate an algorithm that could explain the evolution of the cholera epidemic. Ross introduced parameter calculation that is the measure of variables not directly observed, but whose values can be estimated and verified experimentally. A parameter, defined in terms of probability, measures the likelihood of a certain event.

It can be argued that an epidemiological phenomenon is a dynamic phenomenon. In mathematics, a dynamic application can certainly be studied through the introduction of differential equations, i.e. the solution of these equations provides a variation in time of a given variable.

The time-dependent variable in pandemic prediction applications can be propagation speed (expressed in terms of the first derivative). These variables express the rate of propagation of a disease within a population. However, it is evident that at a certain instant the rate decreases until it stops. In this case, the end of the contagion can be established. Therefore, the growth is limited by a value that depends on the number of vaccinated or cured who developed antibodies. These aspects reduce both the number of people susceptible to infection and the number of those infected.

The solution of a differential equation can certainly be a sigmoid function, which will be limited by the maximum number of individuals in a population and especially by the growth rate (defined in this book as the temperature of the neuron). If the growth rate is greater than zero, the curve will increase, whereas if it tends toward zero, it will indicate the initial case, in which the number of infected patients is known. The growth rate is not a constant value but varies as the pandemic spreads. Using a statistical method of non-linear regression,[48] it is possible to define the rate as a constant value, while this can be also predicted, or the best value can be found using artificial neural networks. We will use the Python libraries to solve a prediction problem, specifically the prediction of the number of COVID-19 cases in the next three months given the training set with patient data.

Note for the reader: this example is not for total beginners at the use of Python or Keras, for a quick simple introduction to the use of Keras Python library for Neural network we provide a link to a useful resources on http://www.ai-shed.com.

In this example Python function libraries are used to resolve a prediction problem, specifically the prediction of the number of COVID-19 cases over three months given the training set of actual data from a public domain database. This example is written in Python 3 and makes use of Neural Network public domain modules. The first code block tells the coder to upload the libraries that the code will need to run. For a description of the different libraries see Chapter 4 of the book.

```
import pandas as pd
import matplotlib.pyplot as plt
from sklearn.preprocessing import StandardScaler
import tensorflow as tf

from tensorflow.keras.models import Sequential
from tensorflow.keras.layers import Dense
from tensorflow.keras.optimizers import Adam
```

The next block of code is used to import the dataset for a public domain database on GitHub. In case of problems check this link for instructions on how to connect to public databases or contact us on http://www.ai-shed.com.

```
'Import the database'
url = 'https://raw.githubusercontent.com/CSSEGISandData/
COVID-19/master/csse_covid_19_data/csse_covid_19_time_
series/time_series_covid19_deaths_global.csv'
n_aff = pd.read_csv(url)
```

Figure 5.33 is an extract of data contained in the databases, with the daily count of affected people per calendar day and per country.

The next block of code is used for data preparation.

The description of each line of code is in the code "comments" below:

```
'Code to retreive the data of the  daily infected'
d = n_aff.loc[:, '1/22/20':]
```

```
'Transpose'
d=d.transpose()

'Sum row-wise'
d=d.sum(axis=1)

'Convert to list data type (to have only numerical
values)'
d=d.to_list()

'Create a new dataset of only two columns '
dataset = pd.DataFrame(columns=['ds', 'y'])

'Lets fill the dataset with values starting at the
 5th column'
dates = list(n_aff.columns[4:])

'Convert dates into date time format'
dates=list(pd.to_datetime(dates))

'fill the dataset with the d and dates we prepared'
dataset['ds']=dates
dataset['y'] = d

'Now we set y as the data snd ds as the index of the
 dataset'
dataset=dataset.set_index('ds')
```

Province/State	Country/Region	Lat	Long	4/3/20	4/4/20	4/5/20
	Afghanistan	33.93911	67.709953	6	7	7
	Albania	41.1533	20.1683	17	20	20
	Algeria	28.0339	1.6596	105	130	152
	Andorra	42.5063	1.5218	16	17	18
	Angola	−11.2027	17.8739	2	2	2
	Antigua and Barbuda	17.0608	−61.7964	0	0	0
	Argentina	−38.4161	−63.6167	39	43	44
	Armenia	40.0691	45.0382	7	7	7
Australian Capital Territory	Australia	−35.4735	149.0124	1	2	2
New South Wales	Australia	−33.8688	151.2093	12	12	16
Northern Territory	Australia	−12.4634	130.8456	0	0	0
Queensland	Australia	−27.4698	153.0251	4	4	4
South Australia	Australia	−34.9285	138.6007	0	0	0
Tasmania	Australia	−42.8821	147.3272	2	2	2
Victoria	Australia	−37.8136	144.9631	7	8	8
Western Australia	Australia	−31.9505	115.8605	2	2	3
	Austria	47.5162	14.5501	168	186	204

Figure 5.33 Input data set.

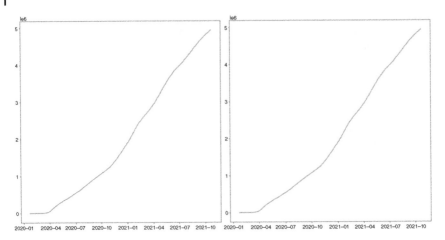

Figure 5.34 Daily cumulative affected people and daily increase rate.

The following piece of code produces two plots (See Figure 5.34).

```
'Plot daily cumulative affected people'
plt.figure(figsize=(8, 8))
plt.plot(dataset*0.3)
plt.savefig('Cummulative daily affected',
 bbox_inches='tight', transparent=False)
plt.show()

'We need to save the daily affection rate (using the
 "diff" function)'
dataset=dataset.diff()
dataset =dataset.loc['2020-01-23':'2022-07-13']

'Plot daily increase rate '
plt.figure(figsize=(8, 8))
plt.plot(dataset.diff())
plt.savefig('Daily affected', bbox_inches='tight',
 transparent=False)
plt.show()
```

Now, it is time to divide the database in training and test set based on the start_date. The values cannot be picked at random because when we train the model on the training set, the purpose is to predict the target values in the future which correspond to date values that are outside of the date values in the training set.

```
def featurize(t):
    X = pd.DataFrame()

    X['day'] = t.index.day
    X['month'] = t.index.month
    X['quarter'] = t.index.quarter
```

```
X['dayofweek']  = t.index.dayofweek
X['dayofyear']  = t.index.dayofyear
X['weekofyear'] = t.index.weekofyear
y = t.y
return X, y
```

```
#print(featurize(dataset))
'Let us divide the database in training and test set based on the
 start_date'
'The values cannot be picked random because when we train the model
 on the training set, '
'the purpose is to predict the target values in the future, '
'which corresponds to date values that are outside of the date
 values in the training set'
```

```
start_date = '2020-07-31'
X_train, y_train = featurize(
    dataset.loc[dataset.index < pd.to_datetime(start_date)])
X_test, y_test = featurize(
    dataset.loc[dataset.index >= pd.to_datetime(start_date)])
```

Standardization of a dataset is a common requirement for many machine learning estimators: they might behave badly if the individual features do not look more or less like standard normally distributed data (e.g. Gaussian with 0 mean and unit variance).

Based on

$$z = \frac{(x - u)}{s}$$

where x is the column before scaling, u is the mean[28] and s is the standard deviation.

```
scaler = StandardScaler()
scaler.fit(X_train)
```

```
scaled_train = scaler.transform(X_train)
scaled_test = scaler.transform(X_test)
```

The following piece of code uses the Keras module to set up and run the neural network, for more information on Keras please consult https://keras.io/guides
Main points:

- Simple feedforward network implemented to Keras Sequential.
- The network has 20 neurons in the input layer, 10, in the hidden layer and 1 output.

```
#Create Sequential model with Dense layers, using the add method
NN_model = Sequential()
```

```
#Dense implements the operation:
NN_model.add(Dense(20, input_shape=(scaled_train.shape[1],)))
#INPUT LAYER as many columns as the input->20 nodes'
```

```
NN_model.add(Dense(10))                          #HIDDEN LAYER
NN_model.add(Dense(1))                           #OUTPUT LAYER

#The compile method configures the model's learning process
NN_model.compile(loss='mean_absolute_error',
 optimizer=Adam(lr=0.001))

#The fit method does the training in batches
NN_model.fit(scaled_train, y_train, validation_data=(scaled_test,
 y_test), epochs=210, verbose=1)

#The predict method applies the trained model to inputs to gener-
ate outputs
NN_prediction = NN_model.predict(scaled_test)
```

The following code plots the chart of the affection rate against the prediction curve (See Figure 5.35).

```
NN_df = pd.DataFrame(NN_prediction)
NN_df.index = y_test.index
plt.figure(figsize=(10, 10))

plt.plot(dataset.tail(60))
plt.plot(NN_df.tail(60))

plt.show()
```

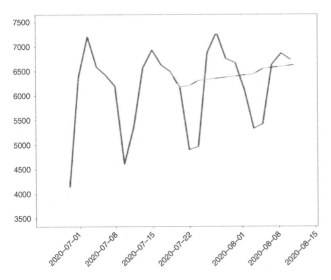

Figure 5.35 Affection rate against prediction (dashed line). The pseudo linear trend between 08-08-2020 and 15-08-2020.

As you can see the prediction is not very good, this demonstrates that either the model requires fine tuning, or that Neural Networks may be entirely the wrong technique to analyze this type of phenomena. This latest aspect is to keep in due consideration in the System Engineering Plan (SEP), in which the related systems engineering process will be applied and adapted to meet the software objectives. The SEP should help software developers to definitely improve their approach to systems engineering rules by providing a well-documented technical report of the software. This planning enables the integration and coordination of technical activities at all levels of management and leads any possible strategy to solve the potential technical issues, met during the software life cycle. The scope of SEP is to ensure proper resource allocation and overall software understanding. The systems engineer helps the developer to integrate the software planning activities into the planning of the system in which the software will operate.

As we have repeatedly written, the Project Manager (PM) gets a key role in the development of every systems engineering process, in that the technical advances achieved to date and the changes resulting from the technical reviews should be considered. In this example, the PM is responsible for the development of the SEP, assisted by the systems engineer and experienced software developers.

5.7 Example 7: Symplex and Game Theory

We always have situations in our lives where we should make a personal decision based on information available and based on how we understand our surroundings. We have never mentioned, however, how to seek solutions in situations where there are interactions between different competitors, where the final output depends on the choices made by others as well. Therefore, it is important to mention the rules which differentiate a competitive relationship from a cooperative one between two or more participants.

Nash had demonstrated that an equilibrium established in a cooperative relationship is not the best result for an individual, but a choice of strategies made by each one to satisfy everyone's profits.

It is necessary to establish that the strategies of an individual depending on individual benefit, are a function of the strategies of the others, dependent on their benefits.

Is there a shared reward? Is there a way to satisfy all participants in competitive environment?

These aspects, which fall within the mathematical discipline defined as game theory, have been treated with high interest not only in economic-financial field, but also in military, political, sociological, psychological sectors, and sports business where the choice of individual strategies and/or a collective is decided based on the earnings received by all participants.

In a challenge between two players, all the utilities, defined as real numbers, are grouped in a single matrix, this allows us to obtain the expected utility if the strategies of each individual player are known. In fact, the choice of strategy for player 1 is based on the choice of a row of the matrix. In contrast for player 2 the choice must be expressed in a column of the same matrix. The intersection of the two strategies asks us what the expected gain of both players will be with a certain probability.

In this example we try to explain the best strategic choice in a **non-cooperative competition**, that is when the decision mechanism concerns a single team that aims to maximize its benefit and choose the best strategy according to the opposing strategies. We build the competition on two teams, the space of their strategies, and the utility functions of each of them.

Defining a **mixed strategy (simultaneous or sequential)** allows us to face an unpredictable opponent in a competition, and therefore to be able to establish that he chooses his own strategy with a certain probability.

In this class of games, introduced by Von Neumann and Morgenstern, players cannot communicate with each other, but each of them chooses the most suitable strategy to maximize their usefulness, considering the rules, conditions of the game and decisions of their opponent.

Is there an equilibrium in a competitive scenario of mixed strategy between two players?

One or more strategies, for each individual player, are used to achieve an equilibrium if no participant can benefit by changing their strategy. Therefore, in an equilibrium, each player's strategy is the best response to the actions of others. Conversely, if a player performs best for himself, regardless of the actions of others, then his strategy will be dominant.

A clarification must be made regarding the cost function which, unlike those used for neural networks, is defined as a linear combination between the utilities acquired for each strategy and the probabilities referred to the choice of each available strategy. The cost function will be both maximized and minimized depending on whether respectively we stand from the observation point of player 1 or player 2.

For the sake of simplicity, the exercise deals with a single utility matrix and the elements of the same matrix are, for instance, positive when referring to the first player's rewards, and negative if referring to the second player's rewards.

How are the utility values of the matrix established? Again, we can rely on expert opinion.

Let us suppose we are in a maintenance department and two firms are competing for signing contracts in a certain period or for a certain number of systems to be maintained. The utility matrix can expressed as the values of costs such as: worked cost/worked time ratios or simply efficiency in terms of time.

The competition must be contextualized to consider not only market demands but also corporate policy choices approved. For example, strategic choices can be established regardless of any competing companies. These aspects could influence future strategic decisions or, in general, could be crucial in a non-cooperative competition with other firms.

A decision to select a certain number of experienced maintenance employees or to focus on the training of their technicians affects the utilities required by the market, as well as determining a change in costs to consider in competition.

Therefore, the assembly of the utility matrix plays a decisive role which also depends on the characteristics of the competitors. Hence, it is necessary to know the challengers and define their main structure.

We can help through a system engineering approach and establish some challenge criteria. For example, if Company 2 offers a maintenance program in terms of efficiency, Company 1's best response could be to reduce costs or vice versa. Finally, assessing dominance of strategies will help simplify our analysis by eliminating dominated or weakly dominated strategies from the game.

Consider the utility matrix A (5 x 5)

We establish that the possible strategies of player 1 are defined on each row of the matrix, while those of player 2 on the columns.

The choice of each strategy is evaluated in terms of probability.

The probability, linearly combined with the corresponding utilities, defines an objective function to be optimized, one for each player.

$$A = \begin{bmatrix} 4 & -1 & -4 & 4 & 2 \\ 0 & -2 & 1 & 4 & -2 \\ -1 & 3 & 3 & -2 & 4 \\ 4 & 0 & 4 & 3 & -4 \\ -4 & 4 & -4 & 0 & -1 \end{bmatrix}$$

Note that the elements of the utility matrix (input values for neural networks) are defined equally for each player. This means that one player cannot achieve much more, or less, than the other, using different rules.

Therefore, the award criteria must be the same for each player. The award criteria must be established clearly and unambiguously at the beginning of each analysis.

For each strategy $\{S_i\}_1$, and $\{S_i\}_2$, arranged to each player, with $i = 1 \ldots 5$, probabilities $\{pi\}_1, \{pi\}_2$, are defined so that objective C will be for each player as follows:

$$C_1 = \{p_i\}_1 \cdot [A] \geq v$$
$$C_2 = \{p_i\}_2 \cdot [A]^T \leq v$$

$$\sum_{i=1}^{5} p_{i1} = 1$$

$$\sum_{i=1}^{5} p_{i2} = 1$$

With $\{pi\}_1, \{pi\}_2$ and v greater than zero.

Both players can choose their strategy to maximize their profit; the best strategy for player 1 is the one which maximizes the minimum utility while for player 2 it is the one that minimizes the maximum utility. These are linear programming problems producing equal solutions. The two problems are the dual of the other, the results of the two problems seen respectively by the first and second player are the same. [Minimax theorem for mixed strategies].

Write the matrix in a spreadsheet

- eliminate dominated strategies
- solve the problem in mixed strategies.

Deleting non-dominant or (weakly) dominant strategies can be useful for finding a point of equilibrium. In general, the mixed strategy that associates to a probability equal to 1 is called a pure strategy. If a pure strategy is a Nash equilibrium in the finite game, then it is also a Nash equilibrium in the mixed strategy game.

Using a spreadsheet in MS Excel, we set the scalar product[36] formula between each row (player 1), and column (player 2) for target cells C.

In the second step we have already found a point of equilibrium (See Figure 5.36), in fact the objectives of the two players are the same. Let us proceed further to

v	p_1	p_2	p_3	p_4	p_5
0.857	0.00	0.48	0.10	0.43	0.00

p_1	0.30
p_2	0.14
p_3	0.56
p_4	0.00
p_5	0.00
v	0.632

$$\begin{bmatrix} 4 & -1 & -4 & 4 & 2 \\ 0 & -2 & 1 & 4 & -2 \\ -1 & 3 & 3 & -2 & 4 \\ -4 & 0 & 4 & -3 & -4 \\ -4 & -4 & -4 & 0 & -1 \end{bmatrix}$$

v	p_2	p_3	p_4
0.857	0.48	0.10	0.43

p_1	0.24
p_2	0.24
p_3	0.52
v	0.857

$$\begin{bmatrix} -1 & -4 & 4 \\ -2 & 1 & 4 \\ 3 & 3 & -2 \end{bmatrix}$$

Figure 5.36 Delete the rows and columns associated with the minimum p.

Figure 5.37 Delete the row associated with the minimum p.

v	p_2	p_4
0.727	0.55	0.45

p_1	0.50
p_2	0.00
p_3	0.50
v	1.00

$$\begin{bmatrix} -1 & 4 \\ -2 & 4 \\ 3 & -2 \end{bmatrix}$$

Figure 5.38 Best strategy between two competitors.

v	p_2	p_4
1.00	0.60	0.40

p_1	0.50
p_3	0.00
v	1.00

$$\begin{bmatrix} -1 & 4 \\ 3 & -2 \end{bmatrix}$$

understand if, by deleting strategy 3 of player 2 which dominates the other two strategies, a possible variation of the choice by player 1 appears (See Figure 5.37).

In fact, player 1 changes his position according to the choice of player 2, focusing only on strategy 1 and 2. We continue to eliminate the dominant strategy.

After the steps taken the results show that player 1 is equally likely to choose between strategy 1 and 3 while player 2 would be more inclined to choose strategy 2 than strategy 4. The objectives for both players are the same, therefore we have reached a point of equilibrium (See Figure 5.38).

5.8 Example 8: Sorting Machine for Lego® Bricks

Source: Daniel West

The web is constellated of thousands of articles on neural networks and during one of those dives I came across an intriguing use for neural networks.

2020 has been an unfortunate year of pandemics and global recession. As many of you, that year, I had lots of "home time" to rekindle with long neglected interests.

Always looking for a constructive use of our time, but also a way to get our minds off from work, we decided to reconnect with our childhood passion for LEGO® bricks, and we rediscovered a forgotten LEGO set in the attic. After assembling it and playing with it we decided to bring this passion to the next level, helping fellow hobbyists to get the missing bricks for their beloved collections. We decided to sell LEGO bricks online.

After several months sorting through crates of assorted bricks, kindly donated by friends and neighbors we realized that sorting them by hand was something excessively time consuming. We concluded that there should be a far better way to

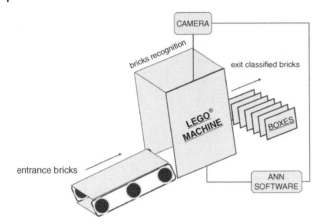

Figure 5.39 LEGO Sorting Machine (for a link to the video or the real machine please visit http://www.ai-shed.com).

identify and sort the parts - after all, manual labor is something we would never accept in our jobs!

With the help from online search engine, we stumbled into a promising project of a young clever talent from Australia: a LEGO sorting machine (See Figure 5.39), built after three years trials and tribulations. A more comprehensive functioning video can be found on our blog.

I was not surprised to learn that, in addition to some clever mechanics, the machine used Neural Networks to accomplish the goal. We would like to add more context onto the complexity of this challenge. LEGO bricks (See Figure 5.40) are made in tens of thousands of variations, each brick is associated with a part number that is made in accordance with specific design rules.

The most arduous part of this hobby is to be able to sort many bricks, not knowing what they are called, which set they come from or even in which color. In the next Figure 5.41 you can see an example inventory.

Figure 5.40 Image by Francis Ray from Pixabay.

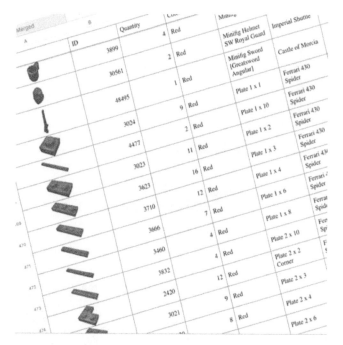

Figure 5.41 What an incredible variety of parts and colors for a simple toy.

After all, it comes across the machine designed and programmed by Daniel West, from now on called "the maker."

His approach was to design an automatic sorter that generates images of all the input components and use mechatronics to have the parts sorted into 18 buckets based on their class. Starting from the architecture of machine domain, shown in Figure 5.42, such is a typical classification problem, and it is achieved using a convolutional neural network programmed in a Raspberry Pi microcomputer[38].

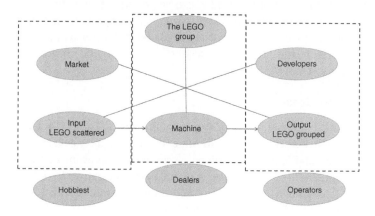

Figure 5.42 Architecture of machine domain.

Figure 5.43   Image by OpenClipart-Vectors from Pixabay.

The key to an accurate classification is the quality of the training set. For parts like LEGO bricks the successful approach was to first try to use real photographs of these compartments. Due to the time consuming nature of manually taking images of the bricks with other factors which may impact the classification (multitude of angles, lighting, textures) the maker decided to use synthetic data, or computer-generated images.

Unfortunately, while this approach is appropriate for many applications (e.g., anomaly detection), it does not work very well in this scenario as the neural network will struggle to understand data that are artificially generated. This is called "reality gap" or a real "cliff" into which to fall (See Figure 5.43). Hence, the solution was domain randomization.

Domain Randomization (DR) is a perfect method to generate data for this king of project. In short, "DR aims to widen the variance of the generated data such that the generated samples encompass the real-life data distribution."

Once Domain randomization is implemented, we have improved the accuracy of our prediction. A final step of fine tuning can be performed by adding some pre-labeled real images (See Figure 5.44) to the training set to improve the accuracy of the classification (See Figure 5.45). This example can be programmed by making use of libraries such as Tensor Flow/Keras/ResNetV2. The synthetic images have been generated using Blender®.

5.8.1   Challenge for Readers

Connect an RGB sensor with an Arduino/Raspberry-Pi board and detect the color of a LEGO brick or other objects. Since RGB values change depending on the orientation of the sensor and lighting conditions, a training set can be created with all RGB values connected to a specific color. A neural network can then be trained to recognize a new, unknown color. A list of RGB values associated with LEGO bricks can be found in the blog connected to the book.

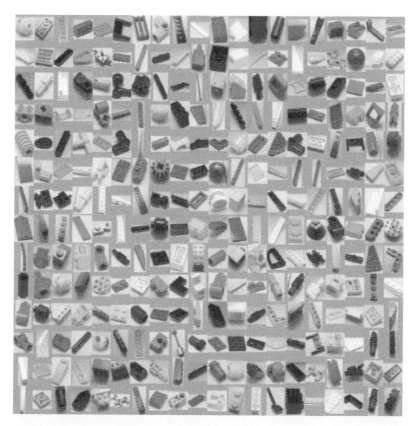

Figure 5.44   Real images.

Figure 5.45   Synthetic images.

Part III

Down to the Basics

6

Input/Output, Hidden Layer and Bias

The probability of A conditional on B is defined as P(A|B) = P(A ∩ B)/P(B),
with P(B) > 0

Definition of Bayes' Theorem

Based on almost all probabilistic approaches, such as Bayes', the explicit thinking, schematized by the possibility tree or by a flow chart diagram, is introduced to express the events through the probabilities of occurrence. On the other hand, the Neural Network gets the ability to produce results without flow chart diagram definition, but the net performs the algorithms well if the inputs, connections between its nodes, and external reinforcements have been correctly represented. Therefore, in this chapter, the Neural Network will be introduced as a paradigm of connections. For those of you that made it this far, a note of respect for your intellectual curiosity. We are at the core of our overview of neural networks, and we will now look at their internal mechanisms. This will help us understand what makes them such useful computational tools for solving problems of classification, prediction, optimisation and more.

Some of our readers might find that the next chapters refer to long lost concepts learnt during their university years. The blog associated with this book will come to your rescue by featuring many articles explaining concepts such as *sum-product matrices*, *bias* and *curve fitting*.

6.1 Input/Output

To build a neural network, we need details on the number of input and output data sets and information on their characteristics. These sets of data will be firstly analyzed and amended to take advantage of other characteristics different from the known ones (*feature extraction*).

Systems Engineering Neural Networks, First Edition. Alessandro Migliaccio and Giovanni Iannone.
© 2023 John Wiley & Sons, Inc. Published 2023 by John Wiley & Sons, Inc.

The number of input neurons is equal to the amount of known data characterizing the problem in question. As we have seen in Chapter 2, this data can be qualitative (such as the color of a flower, the volume of a sound) or quantitative - in both cases we need to turn each element into a real number. To each piece of data, or real number, we associate a neuron n-th, x_n, with n having a value of 1 to N, and N being the total number of neurons, which is equal to the size of input data. Now we know that we can group the N input neurons into one set and, remembering some notions of linear algebra, we can define the following vector[55] $\underline{x}$ $(x_1, x_2, x_n.)$

Finally, we must remember that a single input vector is not sufficient in the learning phase (remember Chapter 2), but we need K input neuron vectors, $\underline{x_k}$ $(x_1, x_2, x_n)_k$, k-th vector, with k having a value of 1 to K with K being the total number of neuron vectors (data set).

The following Figure 6.1 details what has been described previously:

N = total number of neurons, i.e. the total number of input elements per one vector
n = n-th neuron, i.e. n-th element of the input neuron vector
K = total number of samples (*training set*[53] or *validation set* or *test set*[52])
k = k-th sample

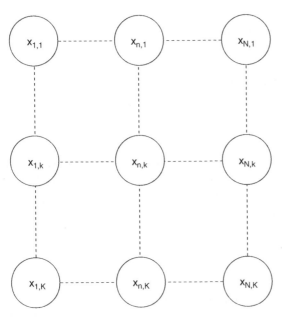

Figure 6.1 Simplified representation of the INPUT matrix of a neural network.

There is a clarification to be made with regards to the input data. In order to convert qualitative data into numerical data, we can define a level (high, medium or low) to which we allocate a real number: 0, 0.5, 1 (this was already presented in parts I and II). When it comes to quantitative data instead, we can use them as is or it might be convenient to convert them for a better representation.

Let us try to understand why a neural network is a very versatile tool.

We are used to describe our experiences by providing many details (variables) of what has happened. These details can often seem not consistent if taken out of context but, when we put them together, their combination leads to a sufficiently clear picture of the experience. A neural network can put together numerous independent variables through a simple process of comparison. In other words, a neural network translates into mathematical formulae what we put into words. Mathematically, the product formula, performed between input and independent variables (i.e. weights), is regularly followed by the neural network, as it becomes clear that the first objective is to find which independent variable is most similar to the input. The mentioned comparison is performed by a neural network in order to find a similar direction and magnitude between input vectors and indipendent vectors (i.e. weights).

Before delving into the true meaning of the comparisons, performed by the neural network, we need to understand the importance of providing data in a form that is easy to process. The normalization[43] helps the network define a correct link among input data regardless of their origin, and quickly reach a process of adaptation. In other words, defining a set of values between 0 and 1 would prevent the network from being influenced by values that are too distant from one another and therefore leading to a trivial solution, mathematically speaking.

Let us say that we want to use our neural networks to predict two types of corrosion that are usually found during inspection of many metallic constructions - superficial and intergranular (a sort of deep corrosion). Each input sample $\{x\}_k$ is made of three elements and one of these will be service hours (how long the structure has been in service). The other two elements of the input vector, such as weather conditions (temperature and humidity variations) and affected area (a few squared centimeters if we are lucky), will not have a direct impact on the final classification of the damage. Service hours, normally measured in the thousands of hours, will largely influence the analysis and therefore the result. This happens when input data have different orders of magnitude.

Neural networks respond well to values between 0 and 1 and to statistical data with an average of 0 and a standard deviation[15] of 1. Here are two simple normalization methods:

$$norm\_x_n^{(k)} = \frac{x_n^{(k)} - avg\left(x_n^{(1)}, x_n^{(K)}\right)}{st.dev\left(x_n^{(1)}, x_n^{(K)}\right)}$$

$$norm\_x_n^{(k)} = \frac{x_n^{(k)} - \min\left(x_n^{(1)}, x_n^{(K)}\right)}{\max\left(x_n^{(1)}, x_n^{(K)}\right) - \min\left(x_n^{(1)}, x_n^{(K)}\right)}$$

As for input neurons, usually the number of output neurons also depends on what the network needs to predict or classify. Output neurons, linked to their respective input neurons, can be qualitative or quantitative and go through the same process as explained previously. We will see how some output data (target) can be processed before entering the network. Some "tricks" help the network to compute faster and obtain satisfactory results.

Figure 6.2 explains well what we have just presented:

J = total number of neurons, i.e. total number or output elements for a single vector
j = *j-th* neuron, i.e. *j-th* element of the output neuron vector
K = total number of samples (learning set, validation set, and test set)
k = *k-th* sample

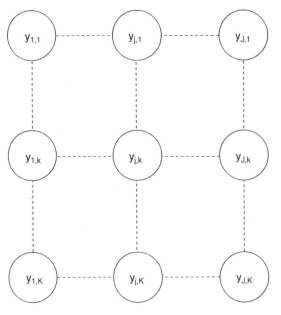

Figure 6.2 Simplified representation of the OUTPUT matrix of a neural network.

Let us now switch our attention to how we manage input and output data. A professional using neural networks usually considers three types of data sets: learning set, validation set and test set (see Chapter 2). To train the network correctly, we need a good number of samples to choose from the whole data set available. By linking the sample input dataset to the respective output set will make the network reach a result that is close to our objective. In general, a prediction problem is specified by a number infinitely many outputs while a classification problem by a small number of possible outputs.

In the following Figure 6.3 a simplified neural network sketch is represented. The input data are associated with the output ones as follows: for each input k-th we have an output k-th; the total number of input neurons N does not necessarily need to be equal to the number of output neurons J.

The **learning** set is used to train the network. These data discern the network's behavior and guide the algorithm toward a validation process.

The **validation** set instead verifies the quality of the network. By elaborating these data, we get to know the degree of accuracy of the proposed model.

The **test** set will be used to test the network with new data, which will ideally come from the same allocation of the validation set. This dataset can also be new but similar to the validation set.

Figure 6.3 Simplified representation of the association between output/input matrices in a neural network for a k-th set of data.

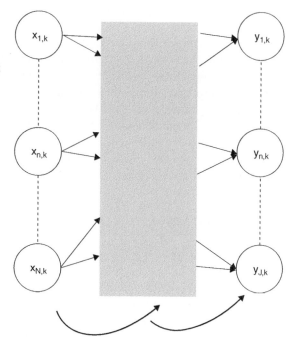

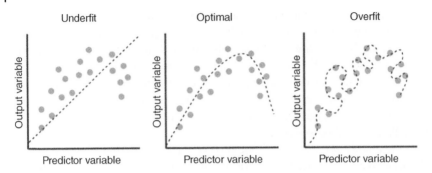

Figure 6.4 Under fitting, over fitting or balanced? (Creative Commons -Attribution -ShareAlike 4.0 -CC-BY-SA 4.0).

Usually, the dimension of the various sets in relation to one another is a topic that would need some time to be discussed, but it suffices to say that the most part of the data is in the learning set, while a small part of them is extracted (and therefore eliminated from the learning set) to become part of the validation set. Around 20% of the data becomes part of the validation set.

This is not a hard rule, it can really depend on the user's decision of how they would like to split the training and validation data set.

In order to understand whether the quality of the network is satisfactory for our final aim, we need to compare the performance of the learning set and the validation set. As shown in the Figure 6.4, if the loss value (see chapter 8) reached with the validation set improves but then degrades, the network is over-fitting - this means that the network has learnt a function that adapts well to the learning data but does not respond well enough to the validation set.

The opposite of over-fitting is under-fitting, which occurs when the performance of the network's training is not satisfactory, and we get high loss value both in the learning set and in the validation set. This can also occur when the training loss value is worse than the validation loss. Our aim is to obtain similar performance in both data sets. In the next pages, where we will present some of the applications, we will explore these techniques in more detail.

6.2 Hidden Layer

So far we have explained that the operations performed by the neural network are linear operations. In fact, in a linear context, adding layers would have no effect because the neural network would always predict the same results. There would always exist a unique correlation between inputs and outputs.

The hidden layer (having specific properties that we will define in the next chapter) is the whole set of neurons between the network's output and input layers. It is possible, in a network, to single out one connection between two general neurons, for example a hidden layer neuron and an output layer neuron. When we are dealing with a feedforward network (moving from left hand to right hand) an input neuron of a single connection it is always specified in the left hand. Such agreement can be extended to general neural network also, where the NN input neurons are on left and NN output neurons are on right (See Figure 6.5).

The presence of hidden layers in neural networks makes them superior to the most part of machine learning algorithms. The higher the number of hidden layers in a neural network, the more time the network will need to process the output and the more complex problems the network will be able to solve.

A supercharged network is not only a waste of encoding efforts and processor resources, but also results in being more susceptible to over training. This is similar to a human being over-thinking a situation: there could be a lot of inactive neurons in hidden levels of the network, as we have mentioned in Chapter 2.

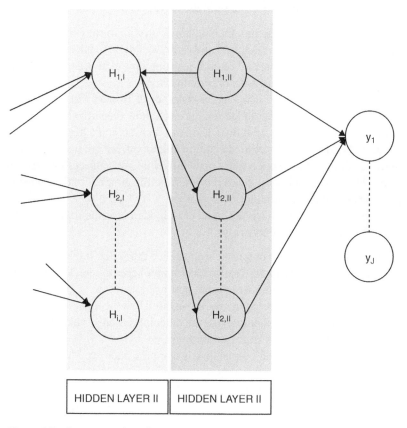

Figure 6.5 Representation of one or more hidden layers.

Hidden layers carry out non-linear transformations of the data introduced to the network. A hidden layer provides special features that improve the NN performances. These performances vary according to the weights (W) associated with them. The following image shows a single hidden layer, and each neuron is associated with a weight (W). Even though each layer produces a specific output, overall, all the layers inside a network contribute to carry out a classification, a regression or clustering in an output layer.

The architecture of hidden layers can be different. As mentioned above, hidden layers can be differentiated by functions applied. For example, a neural network, used for classification of objects, can have a hidden layer that identifies a wheel but not a motorcycle. Nevertheless, if we add layers to identify elements such as the saddle, a metallic chassis, and the headlamp, then the neural network can identify motorcycles based on a set of characteristics.

6.2.1 How Many Hidden Nodes Should we Have?

To find the ideal number of neurons for a hidden layer, with the aim of defining its **dimension**, we need to go through trial and error. We advise to take note of each error and compare them at the end of each learning phase, in order to evaluate which network architecture creates a minimum error.

As previously discussed, too many nodes is not ideal. To start with, it is necessary to have a sufficient number of hidden nodes to allow the network to capture the complexity of the input–output link. The error analysis and a high number of iterations could take us to the desired solution alignment, but we need a reasonable starting point. The network is like a new-born child - we cannot show a series of pictures of animals and ask it to tell the lion from the giraffe. First, we need to teach it how one animal is different from another.

Therefore, in order to establish the number of hidden layer neurons we need, we can apply the following simple rules:

- If the network only has one output neuron and we think that the input–output link is easy enough, then we need to start with a number of hidden level neurons equal to two thirds of the number of input neurons.
- If we have more input and output nodes, or we think that the link between input and output is complex, we should make the hidden layer bigger or equal to the sum of the input and output layers' dimensions.
- If we suspect that the input–output link is complex, then we should set the dimension of the hidden layer with a number of neurons equal or bigger than double the number of input neurons.

When we do not obtain satisfactory results with one hidden layer, we should not increase the number of layers to improve the final result. The addition of a second or more hidden layers will increase the complexity of the code and the processing time.

On the other hand, a network with an extreme processing power and insufficient training data could take us to a highly specific solution rather than a general one. A generalized model will perform better on new input samples. Hornik's theorem[21] also known as universal approximation theorem, affirms that a network with one hidden layer containing a specific number of neurons can round continuously functions on compact convergence topology of R^n. A simple neural network architecture can represent a varied range of interesting functions when they are given appropriate parameters: with enough neurons in a single hidden layer and only one output, we can approximate a continuous function (see Exercise 1, Chapter 5).

In conclusion, in order to improve the algorithm and obtain an appropriate result, we need to explore additional improvements before considering an increase of the number of hidden layers. As mentioned in the previous paragraph, we need to analyze to find a better set of learning data (input and output). Finally, we have to remember that every additional layer may generate more errors in the learning phase, and we could be locked in a layer without an optimal minimum value. Therefore, we would like to highlight the importance of starting with a simple network architecture and considering more than one level, by using a trial-and-error approach.

6.3 Bias

Our brain processes a large amount of information every day, perhaps more than it can actually process correctly. This is why the human brain resorts to alternative ways of processing the information quickly, to save time and energy. There are actual mental alternatives that accommodate these needs exactly. Our brain employs these alternatives to optimize its cognitive processes, but it can still make the wrong decisions for some cases. These expedients are called cognitive biases. It is important to clarify that bias and heuristics are two different things. If cognitive biases are the final error, heuristics are the cause of that error or, we could say, the mental process that takes us to that final error.

As we have mentioned previously, our brain wants to use as little energy as possible and make a decision as fast as possible. Without getting into the details of the topic, see an example to better understand the concept of bias below. At least

Figure 6.6 What influences our choices?. Source: Image by John Hain from Pixabay.

once in our lives, we have found ourselves facing the judgment of new colleagues when starting a new job – they have formed an opinion on us, but not necessarily a negative one. For example, they might think that we are not suitable for the role due to our lack of experience or because we are overqualified.

Neither one of these judgments is correct, or rather, neither one is based on a thorough knowledge of our previous work or academic experience. You can see how we have once again landed on the basics of neural networks: knowledge. As shown in the Figure 6.6, the human understanding of a generic event is influenced by perceptual biases, it is like untangling a maze of uncertainties.

From a mathematical point of view, the bias (scalar) of a neural network, linked to a single neuron or a hidden layer, provides a linear variation of a possibly right value. In the Figure 6.7, the middle line represents the function of all the exact values, while the lines above and below it represent the deviation from all the values on the middle line.

The bias neuron translates the concept of distortion or deviation of the desired output into our network - this is similar to what happens in our brain from a cognitive point of view. A high distortion level means that the model is not "adapting"

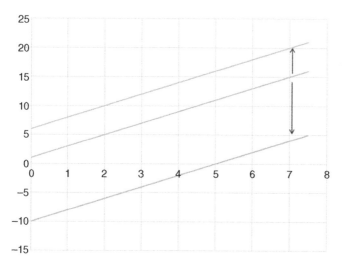

Figure 6.7 Wx + b: how does a (scalar) bias influence the network's neurons?

well to the learning set, and this will result in a considerable learning error. On the other hand, a low distortion level means that the model is well adapted and that the learning error will be small. The bias value has to be as small as possible, as it implies that the network is processing the input data correctly and therefore can provide an evaluation that is as close as possible to reality: we are in the right workplace, but we will need a bit of training!

The bias is initialised by a ± 1 value, so that the function can move higher or lower in the graph. In the chapter on activation functions, we will see how bias values vary to create the needed output values. Even though neural networks can work without bias neurons, in reality these are always added, and their value is part of the general model. Let us use the concept of variance to better understand network distortion.

In statistics, variance is defined as the measure representing how far two sets are from each other. When the variance of a set is small, it shows that the data is generally close to the arithmetic mean. As the variance value becomes bigger, the observations will be more scattered from the arithmetic mean. A high variance value tells us that the model is not able to provide accurate estimations on the validation set.

As shown in the Figure 6.8, the bias can be directly associated with a whole hidden layer, otherwise each bias is associated with a single neuron of a hidden layer, remembering that in the layers following the input layer, the "neural magic" happens.

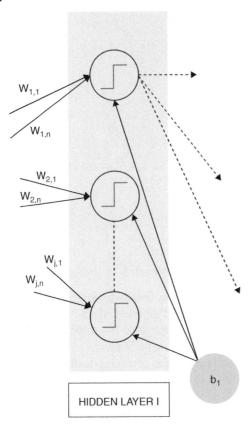

Figure 6.8 Representation of a single Bias associated with the whole hidden layer.

HIDDEN LAYER I

6.4 Final Remarks

Going back to the images of the first paragraph in this chapter, let us add the bias neuron and the concept of sample set to our scheme of things.

Let us consider a *k-th* sample and a generic *j-th* output neuron. Let us link a *j-th* bias to the *j-th* neuron.

In Figure 6.9, you can see a basic outline: N input neurons linked to an output y, which is connected to the bias.

We will continue using the same rules explained in chapter 2 before, and avoid losing track of the process when formulas become more complex - do not worry, it will be simpler than you think!

$$\sum_{n=1}^{N}(W_{jn} \cdot x_n) + b_j = z_j$$

Figure 6.9 Bias connection. Conventional sketch.

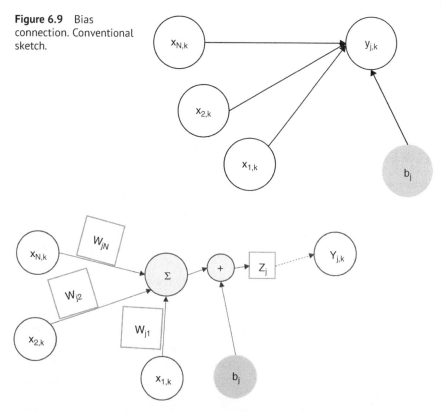

Figure 6.10   Bias represented in a conventional sketch of Neural Network.

You might have noticed that the function z carries the j of the output neuron and of the bias linked to it (See Figure 6.10). All the computed combinations will be found in the output neuron y.

Advancing through the chapters, the mathematical model of a neural network will be more complex, though you will find that the formula we have just provided will be easier to understand. In the next chapter, we will introduce the activation function, one of the last elements that make up a neural network.

6.5   Chapter Summary

In this chapter we started delving into the details of neural networks for solving problems of classification, prediction, optimisation and more - and how they develop their own unique internal mechanisms. There is a clarification to be made with regards to the input data. Neural networks respond well to values between 0 and 1 and to statistical data with an average of 0 and a standard deviation of 1.

To train the network correctly, we need a good number of samples to choose from the whole data set available. The learning set is used to train the network. The validation set verifies the quality of the proposed model. The test set will be used to test the network with new data, which will ideally come from the same allocation of the validation set as the training set.

Over-fitting means that the performance of the network's training is not satisfactory, and we get high loss value both in the learning set and the validation set. The opposite of over-fitting is under-fitting, which means the training loss value is worse than the validation loss.

The architecture of hidden layers can be different. To find the ideal number of neurons for a hidden layer, we need to go through trial and error. Too many nodes are not ideal as they do not capture the complexity of the link between the input–output link. When we do not obtain satisfactory results with one hidden layer, we should not increase the number of layers to improve the final result. The addition of a second or more hidden layers will increase the complexity of the code and processing time.

A simple neural network architecture can represent a varied range of interesting functions. The scalar bias of a neural network provides a linear variation of a possibly right value. A high distortion level means that the model is not "adapting" well to the learning set, and this will result in a considerable learning error. High adaptation level implies that the network is processing the input data correctly.

Questions

1　Why is it worthwhile to normalize the input and output data of a neural network?

2　What are the variables in a neural network? What defines their variation during the training phase of the neural network?

3　What is the meaning of the bias in the neural networks? Can the "mathematical" bias be associated with another meaning (e.g. psychology)?

4　Why do neural networks have hidden layers? What are they used for?

Source

Kahneman D., SLOVIC Stewart Paul, TVERSKY Amos. (2001). *Judgment under uncertainty: Heuristics and biases*, 2001, Cambridge University Press.

7

Activation Function

> *In mathematical physics, quantum field theory and statistical mechanics are characterized by the probability distribution of exp(−βH(x)) where H(x) is a Hamiltonian function*[20] *It is well known in that physical problems are determined by the algebraic structure of H(x). Statistical learning theory can be understood as mathematical physics where the Hamiltonian is a random process defined by the log likelihood ratio function*
> Sumio Watanabe, Algebraic Geometry and Statistical Learning Theory

In Chapter 6, we have established the importance of normalizing the input and output data of a neural network and the need to find a nonlinear operation that causes hidden layers to be sometimes correlated with an input and sometimes not. We have also made it clear that the network's computational algorithms work better if they are processing numbers between 0 and 1. Nevertheless, if we only apply common normalization and summation rules, the optimization will not be defined in all the available mathematical links between the network's neurons.

In order to ensure that the neural network represents a valid mathematical model, we need to introduce some specific functions – activation functions. We will take Particle Physics as an analogy for this. Referring to Fermi statistics, the estimation problem was created to deal with the approximation of some physical quantities that are impossible to calculate otherwise due to their high order of magnitude, especially when singularity is considered.

Estimation is often used in physics, chemistry, engineering, and any other fields where there is a need to understand the physical links between quantities of different nature. Estimations offer a useful way to control the results, and this helps us to identify an error even when calculations become complex. Fermi's approach makes them less susceptible to these errors.

We recommend looking further into this topic and find some reading materials on the origins of these principles, or how they compare to those of Boltzmann and Einstein-Bose. They have answered the following question: how

Systems Engineering Neural Networks, First Edition. Alessandro Migliaccio and Giovanni Iannone.
© 2023 John Wiley & Sons, Inc. Published 2023 by John Wiley & Sons, Inc.

"energetically active" is an elementary particle on average? Fermi first and Dirac afterwards have concluded that, in a system made of a high number of unknown elements (particles) that are in a state of balance and not influenced by external forces, it is more likely that these elements are distributed on different energy levels. The probability of a specific distribution is proportional to the number of ways in which the group of elements can reach such energy levels.

Let us note the Fermi-Dirac equation:

$$n = \frac{1}{1 + e^{FD}}$$

e is a mathematical constant, called the Euler number, and *FD* (in honor of Fermi-Dirac) depends on some chemical–physical quantities of the examined particles. This equation will be relevant later.

We are trying to find a principle to be applied to the elements (neurons) of our system (network), to measure their distribution (output). We believe that the Fermi-Dirac statistics allow us to find the number of weights and biases in a network. Therefore, it is important to carry out another "transformation" of the network data. This is why we introduce some functions, usually non-linear, that accommodate our needs – functions that limit the input (or the sum of all the inputs) between two values.

As we will see, optimization techniques needed to use derivative[9] which allow us to consistently point out a codomain (vertical axis) between 0 and 1. In this chapter, we will go into the details of these "activation functions" that will be subsequently used for our applications:

- sigmoid (logistic function)
- hyperbolic tangent (hyperbolic function)
- ReLU

Logistic functions are generally used as models for population growth over time. Their evolution is exponential until it is limited by a horizontal asymptote.

Hyperbolic functions are "special" functions. They have geometric and analytical properties that are similar to known trigonometric functions (sine, cosine, tangent).

Hyperbolic functions do not depend on an angle but are defined by a set of real numbers.

A fascinating fact on trigonometric functions introduced from a circumference of radius equal to 1: hyperbolic functions are, as a matter of fact, defined by a rectangular hyperbola centered in the origin of the semi axis. The hyperbolic tangent is similar to the sigmoid as its evolution is exponential until it is limited by a horizontal asymptote.

The ReLU function is quite different – the opposite of a linear function, it is equal to zero on half of its domain, where negative values are located. This will cause a phenomenon that we will call "shut down" – half the neurons in the network will have a zero value. The ReLU derivative, different to the one of a linear function, has a codomain equal to 0 or 1. As we will see in the next chapter, linear functions are limited, since the output is proportional to the input. This is why linear functions are not useful in applications where the limit (of the function) helps to inhibit the signal between two or more neurons.

7.1 Types of Activation Functions

Let us establish that the activation function of a neural network is a fundamental tool for reaching deep and systematic learning. Generally, the computing accuracy and efficiency of a network training depends on the type of functions operating on a single neuron. Activation functions have an important effect on the capacity and speed of a neural network's convergence. They could even prevent the network from reaching a solution.

Activation functions are mathematical equations that determine the output of a single neuron in a neural network. Therefore, these functions are linked to each and every neuron of the network (See Figure 7.1), and the input of each neuron will carry weight significantly depending on the accuracy of the model. Activation functions help to normalize neurons' output to an interval between 0 and 1 and −1 and 1.

We would like to share a high level description of common activation functions being used. Normally, a university level knowledge of calculus is needed for this topic, but we think that these functions can also be used as a black box, as you can see in the image above. We are now going to describe the characteristics of activation functions.

The sigmoid function (See Figure 7.2) has in its output (vertical axis) all the real numbers belonging to the interval [0, 1]. This function is not linear, and it pushes its input (horizontal axis) toward the extremes of its output interval. Its application is effective for the classification of two or more separate categories. Its shape is delicate; therefore, its gradient[13] (derivative) is well monitored. The main disadvantage is that the function becomes flat at local extrema. This means

Figure 7.1 Activation functions: why are they important?

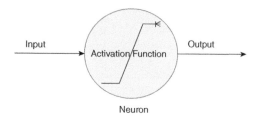

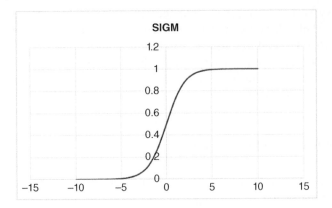

Figure 7.2 Activation function: sigmoid.

that its derivative, or its variation, will be small enough to invalidate the efficacy of the calculation – the computational speed could slow down or stop altogether. The sigmoid function is therefore useful when it is in the final layer of a network, as it helps push the output toward 0 or 1, for example classifying data according to true/false criteria. In general, the sigmoid function is a good choice if the neural network needs to provide only two certain event occurrences (0 and 1). On the other hand, you need to restrict the output to an interval between 0 and 1 in order to predict the likelihood of the event occurrence. For this reason the sigmoid function is often used in the last layer of the neural network, in fact, if it used in previous layers, it could cause uncertainty in some points.

The hyperbolic tangent function (See Figure 7.3) has in its output (vertical axis) all the real numbers belonging to the interval $[-1, 1]$. It is very similar to the sigmoid function, but its derivative (slope) is steeper and restricted in both the positive and negative domains as compared to the derivative of the sigmoid. This will

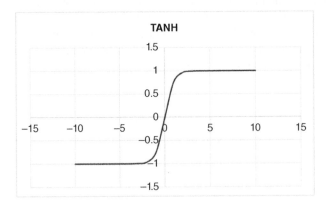

Figure 7.3 Activation function: hyperbolic tangent.

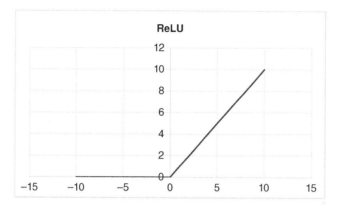

Figure 7.4 Activation function: ReLU (Rectified Linear Unit).

help when calculating the error despite the function output becoming really flat at its extremes.

The ReLU function (See Figure 7.4) is usually divided in two parts. The first has an output (vertical axis) value of zero for input values lower than or equal to zero. The second part instead has a linear trend from 0 to infinity, imposing a slope value if the input value is higher than 0. The search for optimal values will take a long time if we use this function. Also, for uncertain values that we will call fuzzy, we will not be able to find the right distribution in case of great variations. An advantage in using the ReLU function is that the network is lighter, i.e. it can inhibit neurons that were processing values lower than or equal to zero, therefore preventing all neurons from being active at the same time (density issue). The limitation of this function, and of all functions having a flat area and constant values, is that its left side becomes a constant in its codomain. This could again create a gradient – or variation speed – equal to 0, which could prevent that unit from carrying out useful calculations. Calculations are easy and require low energy, therefore ReLU functions are the activation functions that are used the most in the internal layers of the neural network.

7.2 Activation Function Derivatives

All activation functions are required to be continuous and differentiable.[12] The gradient function has to be non-negative and with a local maximum and minimum. In the following images are the three activation functions (full) and their derivatives (dashed).

Let us remind you that:

Sigmoid (can you find any similarities?)

$$a = \frac{1}{(1 + e^{-z})}$$

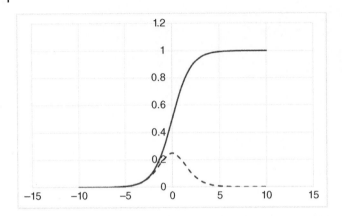

Figure 7.5 The importance of knowing the derivative of each activation function: sigmoid.

and sigmoid derivative (dashed curve in Figure 7.5)

$$a' = \frac{e^{-z}}{(1 + e^{-z})^2}$$

Hyperbolic tangent

$$a = \frac{e^z - e^{-z}}{e^z + e^{-z}}$$

and hyperbolic tangent derivative (dashed curve in Figure 7.6)

$$a' = \frac{4}{(e^z + e^{-z})^2}$$

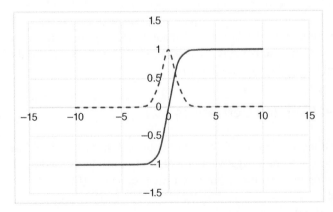

Figure 7.6 The importance of knowing the derivative of each activation function: hyperbolic function.

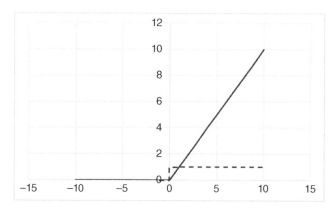

Figure 7.7 The importance of knowing the derivative of each activation function: ReLU.

ReLU

$$a = \max(0, z)$$

and ReLU derivative (dashed curve in Figure 7.7)

$$a = \begin{cases} 0; a = 0 \\ const.value; a = z \end{cases}$$

The gradients (or derivatives) using logistic and hyperbolic tangent activation functions are more limited in the positive domain (horizontal axis) compared to the ones using a ReLU function. This means that the positive section is updated much faster as the training proceeds – based on how weights and bias vary, the derivative of a ReLU function will always produce a stable value in its positive domain, while other activation functions will produce a different value. Vice versa, a null gradient in a negative domain will cause inactive neurons in the network, producing a null output every time negative numbers are processed. The issue of inactive neurons can be solved with new – LU functions – Leaky ReLU, PReLU, and so on. This prevents inactive neurons from interfering with the final result. The presence of inactive neurons is not necessarily negative, we only need to be able to understand which synaptic path is active for a specific application. For example, not all of our neurons will be inactive while we sleep even though the production of melatonin will inhibit some of their functions. Our mind will always be active, maybe processing the solution to a problem from the previous day – the decision we have to make will be clearer upon waking up.

The activation function can be regulated by a "learning parameter" in a different way from another network model. This parameter is usually called "temperature" of artificial neurons. As the performance of the human body is influenced by body temperature, so artificial neurons have a fluctuating processing ability depending

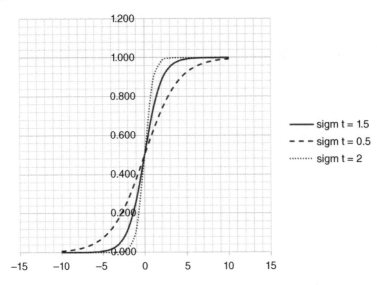

Figure 7.8 Sigmoid function under varying neuron temperature (t = temperature).

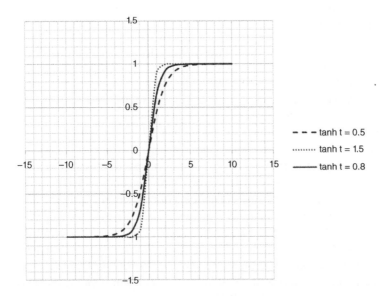

Figure 7.9 Hyperbolic tangent function under varying neuron temperature (t = temperature).

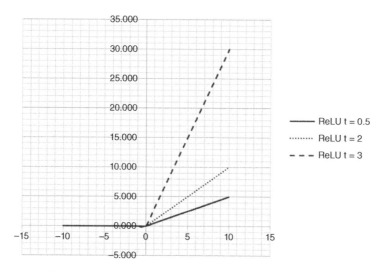

Figure 7.10 ReLU function under varying neuron temperature (t = temperature).

on their temperature. Recent studies have shown the existence of a type of network using a global temperature coefficient that allows the activation function to have a learning variable similar to "synaptic weights". This would allow a high learning speed and would assure more accurate results compared to other networks.

As we can observe in the Figure 7.8 and in the Figure 7.9, as the temperature increases, the sigmoid function and the hyperbolic tangent function can round to a binary threshold [0 1]. On the other hand, the smaller the temperature parameter, the more the function is compressed on the horizontal axis. With regard to the ReLU function instead (See Figure 7.10), we will only have a slope variation and in this case, we will be able to reach a different learning rate[50] during the first phase of the application.

7.3 Activation Functions Response to W and b Variables

Before reading this paragraph, we advise you to get hold of pen and paper. We think it is much easier to understand the constant changes of a neuron by writing this down. In the images below, there may be other responses that activation functions might have apart from the ones proposed above. In this paragraph, we will try to explain how activation functions behave when the parameters of weight (W) and bias (b) vary. Let us extract one neuron from our network, x in input and y in output, and let us assume that the activation function is sigmoid.

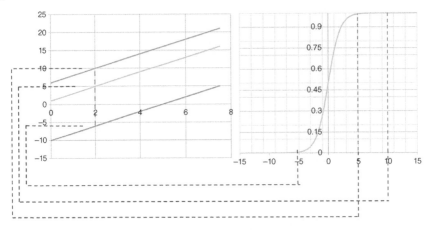

Figure 7.11 Output variation of a neuron based on a varying bias value.

From the Figure 7.11, we notice that the bias variation (the three straight lines) is influencing the output y of the neuron – when the activation function rule is applied, the output value y might not be what we were hoping for. Assuming that the ideal value is 1, or as close as possible to 1, a bias value equal to -10 would give us a value that is instead closer to 0.

Considering the activation function in Figure 7.11, it is clear that an imposed weight ($W = 1$) helps us determine that the function values $z = Wx$ for each input x between 0 and 2 have to be transformed with the activation function in values that are between 0.7 and 1. The aforementioned z values could be affected by the bias that is added to Wx.

This confirms that even a bias variation of $+6$ would not satisfy the behavior of the model influenced by that neuron, as it would always result in outputs y closer to 1, therefore making the neuron inefficient.

Let us say we want to make a decision based on one available piece of data which has a prejudice value equal to a bias value of $+6$ – even if the available data (x) varies, we will always make the same decision in the end.

In the Figure 7.12, we can see that as the weight (W) varies; by applying the activation function rule, the output value can have sudden variations that could be far from the desired value.

For example, with an input equal to 2 on the horizontal axis, and for two weight values of $+2$ and $+3$, the output value on the vertical axis will be close to 1. Vice versa, if the weight value were -3, the output value would be closer to 0. It is important to notice the response of each neuron in the context as a network. This will confirm that the variation of a single weight does not depend only on the desired output – be it 0, 1 or a value in between – but it also depends on the weight variation of other connections.

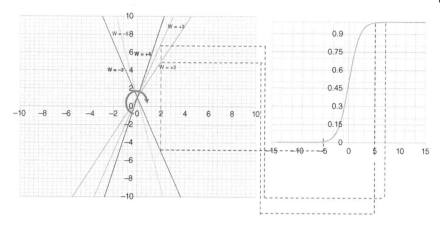

Figure 7.12 Output variation of a neuron based on a varying weight value.

Their combination on each of the neurons, linear $(Wx + b)$ at first, will produce a balance on the entire network that depends on all the weights and biases – similar to a scale with many arms. As illustrated by the Figure 7.13, we consider that the neural network can be represented by a little man striving to find a balance among the many independent variables in order to correlate the many other variables depending to a given event.

Figure 7.13 Weight and bias values depend on each other.

7.4 Final Remarks

As previously done, we add the activation function (See Figure 7.14) to the connection derived from the network and we apply the rule considering a *k-th* sample, an *n-th* input neuron and an *j-th* output neuron. We also associate a *j-th* bias to the *j-th* neuron, remembering that all the written combinations will be found in the output neuron y.

It is probably easier to remember all combinations defined on one or more connections by using the following Figure 7.15. The reader is free to use the representation that they find clearer, as long as the operations are carried out correctly.

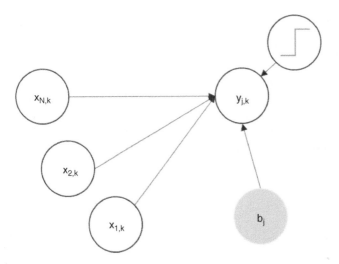

Figure 7.14 Introducing an activation function. Conventional sketch.

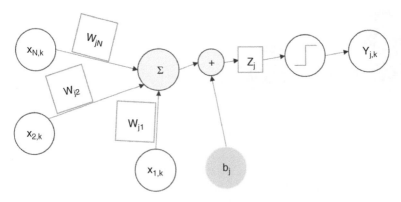

Figure 7.15 NN representation. Conventional sketch.

Remembering that z is the linear combination between weights and input neuron values added to bias:

$$\sum_{n=1}^{N}(W_{jn} \cdot x_n) + b_j = z_j$$

we apply the activation function, for example sigmoid, and we get to the definition of the function y.

$$y_j = \frac{1}{1 + e^{\sum_{n=1}^{N}(W_{jn} \cdot x_n) + b_j}} = \frac{1}{1 + e^{z_j}}$$

Obviously, we can apply whichever function we think is most suitable for our goals, as the neural network's fine-tuning process is iterative. In the meantime, the reader will be able to practice on finding the network output by applying other activation functions.

NOTE: For further reading, we shall mention that some activation functions are not selected in the same way we have done in this chapter but would depend on a casual distribution of input values (stochastic functions).

7.5 Chapter Summary

Activation functions are mathematical equations that determine the output of a single neuron in a neural network. These functions are linked to each and every neuron of the network, and the input of each neuron will carry weight depending on the accuracy of the model. The sigmoid function has in its output all the real numbers belonging to the interval $[0, 1]$. The hyperbolic tangent function has in its output (vertical axis) all the real numbers belonging to the interval $[-1, 1]$. It is very similar to the sigmoid function, but its derivative (slope) is steeper and restricted in both the positive and negative domains.

Gradients using ReLU and hyperbolic tangent activation functions are more limited in the positive domain (horizontal axis) This means that the positive section is updated much faster as the training proceeds. The presence of inactive neurons is not necessarily negative, we only need to understand which path is active for a specific application. On the other hand, the smaller the temperature parameter, the more the function is compressed on the horizontal axis.

Questions

1   What are the activation functions for? How do their characteristics affect the bias and weight variations?

2   Why must the neural network perform nonlinear operations?

3 What are the advantages of using the ReLU function?

4 What is meant by balance "weight-bias"?

5 Seek activation functions other than those mentioned in the text and try to describe their characteristics.

Source

Kriesel, D. (2007). A Brief Introduction of Neural Network (online resource). http://www.ai-shed.com (accessed 4/02/2021).

8

Cost Function, Back-Propagation and Other Iterative Methods

> *Dimensional models implemented in relational database management sys-*
> *tems are referred to as star schemas because of their resemblance to a star-like*
> *structure. Dimensional models implemented in multidimensional database*
> *environments are referred to as online analytical processing (OLAP)[…]. Both*
> *stars and cubes have a common logical design with recognizable dimensions;*
> *however, the physical implementation differs. When data is loaded into an*
> *OLAP cube, it is stored and indexed using formats and techniques that are*
> *designed for dimensional data. Performance aggregations or precalculated*
> *summary tables are often created and managed by the OLAP cube engine.*
> *Consequently, cubes deliver superior query performance because of the precal-*
> *culations, indexing strategies, and other optimizations […]. The downside is*
> *that you pay a load performance price for these capabilities, especially with*
> *large data sets.*
>
> Ralph Kimball, The Data Warehouse Toolkit: The Definitive
> Guide to Dimensional Modelling

In this chapter, we will introduce loss or cost functions allowing us to measure the error of the network while it is evaluating, classifying or predicting a specific event starting from a set of data.

Using a mathematical approach, we will find the minimum value that these functions can reach and define the network's ability to get as close to reality as possible - the smaller the error, the better the ability of the network to describe a specific phenomenon. Continuing to use a mathematical approach to explain this concept, we only need to define a function of n variables (weights and bias) and study its minimum value. There are many books on multivariable functions, and surely it would be helpful to understand the problem we are studying in case we wished to look at it from a topological, algebraic, and geometric point of view. Therefore, we will not provide a mathematical explanation of a multivariable function, but we will provide some tools for the reader to understand the notions

Systems Engineering Neural Networks, First Edition. Alessandro Migliaccio and Giovanni Iannone.
© 2023 John Wiley & Sons, Inc. Published 2023 by John Wiley & Sons, Inc.

at the basis of back-propagation. In general, back-propagation is the technique by which we send an error signal backwards into the network for one or more hidden layers, therefore reducing the error. We obtain this by varying bias and weights values, and the derivative of the activation function.

We hope that, at this point, we have succeeded in transferring the concept: operating in complex systems, nonlinear optimization problems could be solved. Aiming to solve prediction and classification problems, neural networks model nonlinear problems. The performance of the neural model can be measured by a function E, which expresses the reliability of the model and, in mathematical terms, a dependency with the independent variables (weights and biases).

$$E(\boldsymbol{w}, \boldsymbol{b}) = f(\boldsymbol{w} \cdot \boldsymbol{x} + \boldsymbol{b})$$

A simple quadratic function is represented in Figure 8.1, so as to understand the usefulness of the derivation that will speed up the calculus during the training phase. In any case, the complexity of the problem increases dramatically when the function E depends on two or more independent variables ($w_1, w_2, \ldots b$). The derivatives of the function E won't be scalars but they become a column vector and a matrix, respectively (gradient - first partial derivatives and "Hessian" matrix - second partial derivatives).

Therefore, when the function E, as shown in the figure, has only one minimum, then the learning phase speeds up considerably. In the nonlinear problems, the error function is not always convex, so the speed of calculation increases. As a relevant example to understand the topic we will present finding the shorter distance between two points in a space of n dimensions or, in engineering terms, breaking down a system of n degrees of freedom. We define dimensions as a list of numbers representing the coordinates and we need to describe the status of the system (neural network). In other words, we need three coordinates to define the position of a point in space, while we need n coordinates in a space with n dimensions.

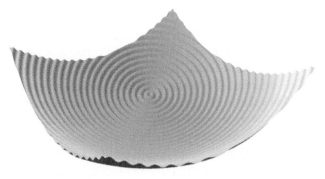

Figure 8.1 Example of a function of two variables.

Before continuing with this chapter, we recommend you look into the following concepts:

- derivatives;
- gradient;
- linear and nonlinear function;
- monotonic and non-monotonic function;
- convex and concave functions[27]
- linear algebra.

To help the reader, we have added a list of useful resources on the blog associated with the book.

The easier representation of cost is mean squared error, used in most applications presented in this book – generally the error is equal to prediction value minus target value. The target is reaching the real value or known result, while the prediction is the result obtained by the neural network.

An important part of neural network training is to find the minimum of mean squared error function, also known as loss function L2 – this also means calculating the derivative of the function and matching it to zero. The expert reader should read on the topic of the first derivative of function with n variables. In our case, it would be correct to call it the gradient function or a vector of partial first derivatives[10] Minimizing a function, or finding its minimum value, is not straightforward, and such estimation leads to:

- Severe non-linearity of the error, creating steep and flat areas on the function surface.
- High values for weights and bias on a high number of samples.
- Presence of local minima (relative).

On this topic, it is important to search for valid iterative methods to calculate the variables to the problem (weights and bias). These methods are divided into two different classes, and to each we associate a specific method - we advise you start delving into the ones you understand better:

- **Batch methods** in which weights and bias are updated by using information from all the samples in the training set.
- **On-line methods** in which weights are updated cyclically only taking one sample into account.

In this book, we only use batch methods. These allow a quick reduction of the total error of the network compared to the initial estimate. We also think that these are more practical methods for those using neural networks for the first time.

In general, the most appropriate methods for creating training algorithms are the ones requiring only the first derivatives and that can be used in problems with big dimensions, i.e. where the number of variables is very high. The methods we will describe in this chapter, apart from back-propagation, are the gradient method, the conjugate gradient method, and Newton's method.

8.1 What Is the Difference between Loss and Cost?

Loss and cost are synonyms when it comes to neural networks. Nevertheless, we have to clarify that cost and loss are the key to error optimisation. Error is also a mathematical function and as such it can be analyzed, studied and manipulated. From a neural network point of view, loss and cost are referred to in different ways:

- maximizing probability in retrospect (e.g. naive Bayes);
- maximizing a fitness function (genetic programming);
- maximizing the compensation function/total value (backup learning);
- maximizing the intake of information/reducing the impurities of the node (CART decision tree classification);
- minimizing the cost (or loss) function of the mean square error (CART, decision tree regression, linear regression[47], linear adaptive neurons, and so on);
- maximizing logarithmic probability or reducing the loss (or cost) function of cross entropy;
- reducing hinge loss (support vector machine).

The expert reader can rely on various written materials to shed some light on the concepts mentioned above, as we will only cover them partially in this chapter. Please also have a look at our reference section.

Some common loss functions are presented below, where y_i is the *i-th* result of the model and $\hat{y}_i$ is the *i-th* true value (target).

Regression loss functions:

- Mean Squared Error Loss

$$MSE = \frac{1}{n} \sum_{i=1}^{n} (y_i - \hat{y}_i)^2$$

- Root Mean Squared Error

$$RMSE = \sqrt{\frac{\sum_{i=1}^{n} (y_i - \hat{y}_i)^2}{N}}$$

Where y_i is the *i-th* result of the model and $\hat{y}_i$ is the *i-th* true value (target).

- Mean Absolute Error Loss

$$MAE = \frac{1}{n} \sum_{i=1}^{n} |y_i - \hat{y}_i|$$

Binary Loss Functions (or for binary classification):

- Binary Cross-Entropy

$$L = -\frac{1}{n} \sum_{i=1}^{n} y_i \cdot log(\hat{y}_i)$$

- Hinge Loss

$$L = max(0, 1 - y_i \hat{y}_i)$$

By using the representation of a cost (or loss) function, we understand how the neural network finds an optimal solution to a problem. As with every function, it is possible to trace a geometric shape to represent it. Studies on the geometry of cost and loss functions can provide an in-depth analysis toward the understanding of reality as seen by computers (computer vision).

Euclid formulated the main rules allowing us to analyze 3D and 2D images that are perceivable by human senses. However, these rules do not explain how the brain interacts with them. Even though most mathematical thinking is based on the concepts we have been taught in school since childhood, Euclidean geometry does not provide enough means to define the representation of multivariable functions.

We will leave it to the reader to find more sources to explore the world of multivariable functions and their graphic representation. We can recommend works written by Riemann, Gauss, Lagrange, Cauchy, Banach, Hilbert, just to mention a few. This will help you find the basis of a geometry that goes beyond the old Euclidean rules which, apart from the 5th postulate, define a curved geometry, not limited to three dimensions. With regards to neural networks, we recommend further reading on Pinkus' theorems on the existence of continuous functions of three variables - as non-comparable to combinations of continuous functions of two variables (see Hilbert). It is proved that multivariate continuous functions can be represented and simplified into overlapping one-variable functions (Kolmogorov). A probabilistic approach pushes us toward an almost certain definition of the event by using mathematical series and sequences. **We can state that neural networks help us fill the gap between random phenomena and the complete knowledge of the status of system variables.**

Finally, we would like to highlight that a (non-linear) link between quantities supposedly having nothing in common with each other is formed during the optimization of the error function (non-monotonic function). Once this link is found, we can evaluate the efficiency of the algorithms used to optimize the error, and we are able to associate our network to a specific application.

For example, after discussing our work experience and qualifications during a job interview (See Figure 8.2), we are usually asked about our personal interests or leisure activities. An inattentive reader might think that there is no link between

Figure 8.2 Job interview as error optimization algorithm.

these two aspects, or that they do not belong to the context of a job interview. However, the evaluation will be more efficient with more variables the interviewer can put together, as these will provide the best overall picture of a candidate. A precise evaluation would make it easier to find the best role for the candidate or to find reasons as to why they will not be considered (constructive feedback[7]).

8.2 Training the Neural Network

As mentioned in the previous chapters, training consists of input–output pairs (x, y) and, for each pair, a real value that represents the ideal output (*target*). The trained network must provide a correct output to each input. According to the definition of supervised learning[5] the network's parameters (weights and bias) have to be determined with relation to the outputs in the relevant training samples, calculating the error between the output value and the one desired for each training set sample (generalization). Future revisions of this book will probably include the concepts of unsupervised training, in which not only the output samples are unknown, but the network's parameters are determined with clustering techniques applied to the input samples.

From now on, we will assume the error is a continuous and differentiable function (this takes us back to the concept of derivative!) as our network and the minimization methods require derivative calculations. Addressing minimization theoretically can bring inadequate results, in addition we need to use suitable intuitive methods to choose and define the training, validation and test set.

In general, we follow four strategies when it comes to building multilayer networks:

- **structural stability**: a number of units is chosen through the training of a sequence of networks in which the number of neurons is increased (or decreased). For each of these networks, parameters are defined by minimizing the error on the training set, and the performances of the networks are compared through cross-validation techniques - evaluating the error that each network creates on another set of data (validation set) not included in the training one. The selected network is the one providing the minimum error on the validation set. The performance of a trained network is evaluated by using a third set of data the test set - which does not interfere with the network's architecture or with the determination of the network's parameters.
- **Regularization technique**: this is based on penalizing the error function with the consequence of tightening the set from which parameters are chosen (weights and bias). The training is performed by defining a new loss function which includes a penalty depending on the data that has not been evaluated in the training phase. For example, if we wanted to analyze the spread of a specific epidemic or pandemic, the increase or decrease would depend on a new input parameter: "adherence to prevention guidelines" that have not been added to the input data. Therefore, it is possible to introduce a factor that determines the error trend based on an input parameter, such as a prevention guideline, which can be difficult to evaluate at the beginning of the spread.
- **Early stopping**: As an alternative to the regularization technique, we can prematurely stop the minimization of the error function. This technique is based on the idea of evaluating the network's error on the validation set periodically and during the minimization process. In the first iterations, the error on the validation set can decrease with the loss function, and it can increase if the error on the training set becomes small enough. The training process ends when the error on the validation set starts to increase, because this can highlight the start of a network's overfitting phase - the phase in which the network tends to fit the training data at the expense of generalization.
- **Data increase**: the more training data, the better the network performance. Nevertheless, it is not always possible to obtain more data. We can instead increase the amount of existing data by creating their artificial variations. In the case of images, we can apply rotations, translations, clippings and other techniques to produce new variations.

There are various training laws, and most of them are a variation of the most popular one - Hebb's rule. Latest research tends toward modeling based on biological training or instead toward human perception adaptation. We highlight again how human knowledge is fundamental to obtain an ideal neural performance. This clearly shows that training is more complex than the simplified mathematical laws currently used.

8.3 Back-Propagation (BP)

As mentioned, an iterative method such as in Python applications is the back-propagation method. BP as a term does not leave room for interpretation, and it refers to the technique of proceeding through the network, opposite to the feedforward direction - starting from the output, we will return to the input neurons (See Figure 8.3).

This is one of the most popular training methods, and it makes it possible to define the training algorithms for multilayer networks. For this purpose, we will represent a generic network outline made out of a *n-th* neuron in the input layer, a *j-th* neuron in the output layer, and a *i-th* neuron in a hidden layer.

For a simpler and more intuitive explanation of the way back-propagation works, we recommend the interactive resources by Grant Sanderson, mentioned in the references.

Before proceeding any further, we remind you that calculating the derivatives of error functions is key for this method as well as for all other iterative methods. Let us imagine a simple network (See Figure 8.4) where each layer is made of one neuron only, with the notation used in the previous chapters, and ΔE as the overall cost of the network.

This process is based on updating weights and bias backwards.

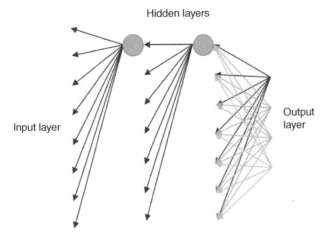

Hidden layers

Input layer

Output layer

Figure 8.3 The network is in auto correct, back-propagation (simplified graph).

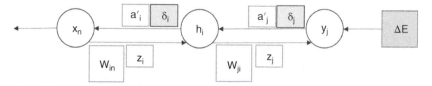

Figure 8.4 Neural Network flow against Back Propagation flow.

Starting from classic iteration formulae:

$$\{w_{ji}\}^{e+1} = \{w_{ji}\}^e + \{\Delta w_{ji}\}^e$$
$$\{b_j\}^{e+1} = \{b_j\}^e + \{\Delta b_j\}^e$$

In the following explanation, we will only show the update of weight W, but it is possible to proceed in the same way for updating the bias.

Weight variation, i.e. cost susceptibility to small weight variations, is calculated with the following:

$$\Delta w_{ji} = -\eta \frac{\partial E}{\partial w_{ji}}$$

The learning rate η is normally gradually adjusted because the convergence (error goes to zero) could happen relatively quickly when the error variation is small.

Let us split the partial derivative of the error function against the weight of the general link i-j (also known as chain rule):

$$\frac{\partial E}{\partial w_{ji}} = \frac{\partial E}{\partial z_j} \cdot \frac{\partial z_j}{\partial w_{ji}}$$

and by defining:

$$\frac{\partial z_j}{\partial w_{ji}} = h_i$$
$$\frac{\partial E}{\partial z_j} = \delta_j$$

it follows that the error variation in relation to the weight (or cost susceptibility to small variations of weights or bias) can easily be written as:

$$\frac{\partial E}{\partial w_{ji}} = \delta_j \cdot h_i$$

Now we only need to calculate the delta parameter of the j-th output neuron, while h_i is the output of the activation function of the i-th neuron in the hidden layer.

$$\delta_j = \frac{\partial E}{\partial z_j} = \frac{\partial E}{\partial y_j} \cdot \frac{\partial y_j}{\partial z_j}$$

where:

$$\frac{\partial y_j}{\partial z_j} = a'_j$$

is the derivative of the activation function of the output layer neurons, while the error variation in relation to the output y is easily calculated by deriving the cost function, when differentiable:

$$\delta_j = \frac{\partial E}{\partial y_j} \cdot a'_j$$

with regard to the second part, from the hidden neuron to the input neuron, we split the error δ referred to the hidden neuron and therefore the combination of weights entering the hidden neuron in feedforward:

$$\delta_i = \frac{\partial E}{\partial z_i} = \frac{\partial E}{\partial z_j} \cdot \frac{\partial z_j}{\partial z_i}$$

δ parameters are the only ones to have an influence on the calculation of weights on every link, as they allow to overcome the possible lack of information in the hidden layers. This means that we can find a relation between activation functions and connection weights, regardless of the output and input of each neuron. Therefore, remembering that h (as y) is the output of the activation function of the hidden neuron, it follows that:

$$\frac{\partial z_j}{\partial z_i} = \frac{\partial z_j}{\partial h_i} \cdot \frac{\partial h_i}{\partial z_i}$$

$$\frac{\partial z_j}{\partial z_i} = a'_i \cdot w_{ji}$$

$$\delta_i = \delta_j \cdot a'_i \cdot w_{ji}$$

in conclusion, if we think that each neuron is connected to multiple j neurons, the delta error of the i-th neuron is:

$$\delta_i = a'_i \cdot \sum_{j=1}^{J} \delta_j \cdot w_{ji}$$

We now only need the error variation in relation to w_{in} weight, that is to say the weight variation in the generic connection n-i between the hidden layer and the input layer of the network. We apply the same relation that we have used for the output layer:

$$\Delta w_{in} = -\eta \frac{\partial E}{\partial w_{in}}$$

where again

$$\frac{\partial E}{\partial w_{in}} = \frac{\partial E}{\partial z_i} \cdot \frac{\partial z_i}{\partial w_{in}} = \delta_i \cdot x_n$$

In conclusion, to apply the BP method, we have to first calculate the output of the network in feedforward starting from the input vectors, and then calculate the δ parameters of each layer backwards, according to the formulae introduced previously. For each weight update, we check the total error of the network. This weight - and bias - update process can be stopped when the error is close to zero at the *e-th* iteration (epoch). This means that we have reached convergence, i.e. our network is trained.

The iterative process can be stopped when the error is small enough for our goals.

If we wished to update weights by applying the BP method online, then we would have to take one input vector at a time, that is to say one training set sample at a time. Following each iteration, we set apart the error that needs to be added to all the errors calculated with other samples of the training set, and then check at the end of every cycle that the total error tends toward zero (convergence).

In the case of a BP batch method instead, we consider a total error function - we evaluate all the input samples at the same time, so the error variation in relation to the weight considers the total variation of the network error.

8.4   One More Thing: Gradient Method and Conjugate Gradient Method

If you have come to understand the back-propagation method, then you will find the gradient method to be straightforward. We will use simple examples to explain this method and it will be the reader's task to make it work for neural networks.

Firstly, let us introduce two new parameters. Direction d (vector), which we will cover when we are moving along the error function, and step $alpha$ (scalar) which defines the variation of d.

The gradient method is characterized by a search of each step. Imagine going from A to B and reaching B step by step - we can do this by setting a constant value for the step or a variable one, while studying the error function. Let e be the iteration and we start from a set value of the weight. To simplify, we will only show updating the weight W- and we have touched upon in Part II that weight and bias values are chosen initially at random:

$$\{w^{e+1}\} = \{w^e\} + \alpha^e \cdot \{d^e\}$$

We remind you that we are presenting iterative methods, therefore the various steps have to be repeated as many times as it takes to find the minima of the error function, and that we define a gradient (vector) when functions have one or more variables. In practice, our network will have more than one weight and one bias, otherwise we should have a connection between only two neurons!

$$\{d^e\} = -\{\nabla E(w^e)\}$$

The limitation of this method is the appropriate choice of a constant step: if it is too small, the iterations will be many and will jeopardize the speed of calculation; if too large, we might end up being inaccurate as we would go around the solution without ever finding the right value. On this note, there are various techniques of one-dimensional searches to determine the step, which would guarantee the convergence and improve the behavior of the method. These techniques are based on the definition of "intervals of values" for the step so that we can guarantee:

- A sufficient movement.
- Convergence of the objective function.

Here is a step formula, but there are more available in the literature:

$$\alpha^e = -\frac{\{\nabla E(w^e)\}^T \cdot \{d^e\}}{\{d^e\}^T \cdot [\nabla^2 E(w^e)] \cdot \{d^e\}}$$

Let us assume that we have to reach point B (objective: error function minima) starting from point A in the shortest time possible. If we also assume that all directions chosen during the movement are small segments of various lengths (one-dimensional approach), we can choose the quickest way if we are able to select the next segment and modify the direction at the same time (See Figure 8.5).

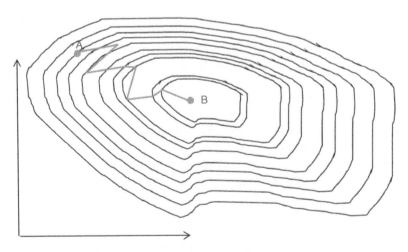

Figure 8.5 Find the solution by the gradient method.

We can obtain this by modifying the search direction and adding a factor that is dependent on the direction established previously. Generally, progressing toward the conjugate direction is the goal of obtaining a quicker convergence compared to the gradient method, especially without using second derivatives and complex matrix operations. This method generates an equivalence relation that represents the same linear application and it converges to the ideal solution by a number of iterations smaller than or equal to the problem's dimensions. The step is determined by means of one-dimensional search.

Let us consider the error function:

$$E(w_1, w_2) = w_1{}^2 + w_2{}^2$$

from which we can easily get to the minimum value (0,0). We want to reach the same minimum value by using the conjugate gradient method. With $e = 0$, we imagine that the starting point is:

$$\begin{Bmatrix} w_1 \\ w_2 \end{Bmatrix}^0 = \begin{Bmatrix} 2 \\ 2 \end{Bmatrix}$$

At step 0:

$$\{d^0\} = -\{\nabla E(w^0)\} = \begin{Bmatrix} -4 \\ -4 \end{Bmatrix}$$

$$\alpha^0 = -\frac{\{\nabla E(w^0)\}^T \cdot \{d^0\}}{\{d^0\}^T \cdot [\nabla^2 E(w^0)] \cdot \{d^0\}} = 0.5$$

it follows that the starting point for the following step is:

$$\begin{Bmatrix} w_1 \\ w_2 \end{Bmatrix}^1 = \begin{Bmatrix} 2 \\ 2 \end{Bmatrix} + 0.5 \cdot \begin{Bmatrix} -4 \\ -4 \end{Bmatrix} = \begin{Bmatrix} 0 \\ 0 \end{Bmatrix}$$

Proceeding with the iterations, we will see how the point obtained after the first step is the error function minima, and the analyzed function is only an easy example to understand how to apply the iterative gradient method.

8.5 One More Thing: Newton's Method

The "Pure" Newton's method defines another formula for the direction d by introducing the concept of second order derivative (derivative of the gradient). This method is convenient if the functions we have to minimize are quadratic and convex, as they will allow us to have a fast convergence.

$$\{d^e\} = -[\nabla^2 E(w^e)]^{-1} \cdot \nabla E(w^e)$$

In other words, we could get to a solution - B in the image - in one step, starting from A (See Figure 8.6).

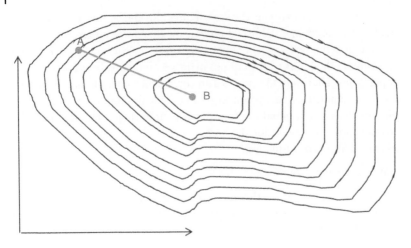

Figure 8.6 Find the solution by Newton's method.

Here is the error function:

$$E(w_1, w_2) = w_1{}^2 + w_2{}^2$$

starting from the initial point:

$$\left\{ \begin{array}{c} w_1 \\ w_2 \end{array} \right\}^0 = \left\{ \begin{array}{c} 2 \\ 2 \end{array} \right\}$$

we will obtain below by one step:

$$\{d^0\} = \left\{ \begin{array}{c} 2 \\ 2 \end{array} \right\}$$

so, given that:

$$\{w^{e+1}\} = \{w^e\} + \alpha^e \cdot \{d^e\}$$

and that in Newton's pure method the step is not definite, it follows that:

$$\left\{ \begin{array}{c} w_1 \\ w_2 \end{array} \right\}^1 = \left\{ \begin{array}{c} 0 \\ 0 \end{array} \right\}$$

As seen in the previous paragraph, we would be able to find the minima of error function even with this approach.

You may have guessed by now that the error functions used for neural networks are more complex than the ones in this example, but the underlying principle should be clear.

8.6   Chapter Summary

- The total error of the network is calculated with the feedforward process.
- The cost function measures the difference between the (known) objective and the output value of the neural network.
- The optimization of the prediction process happens during the training phase.
- It is critical to choose the right cost function to obtain the desired result when it comes to problems of machine learning. Two models that are measuring the same event could produce different results using 2 different computing techniques.
- Backpropagation will happen with the feed-forward process in tandem during the training phase, as the error value for each neuron is a function of the variation of the total error of the network.
- Structural stabilization can be achieved where the performances of the networks built with the different number of neurons are compared by a technique of cross-validation.
- Regularization is a way of adding a penalty to the error function. The network's training is executed by defining a new objective function, which includes a penalty depending on the uncertainty of the event we observe.
- We can stop the minimization of the error function ahead of time using the early stopping technique. The training process stops when the error on the validation set starts increasing.
- We can increase the amount of existing data by generating artificial variations.
- Gradient and Newton's methods help find the point of convergence (error minimization) in the shortest time possible (epoch).

Questions

1   What is the difference between supervised learning and unsupervised learning?

2   When is the batch method used?

3   What are the loss and cost functions for?

4   What is the difference between loss and cost functions?

5   What is the "error" when we talk about neural networks? How is this error reduced?

Sources

Graupe D. (2007). *Principles of Artificial Neural Networks (2nd Edition) Advanced Series on Circuits and Systems.*Vol. 6, World Scientific Publishing Co. Pvt. Ltd.

Mulla Z. (2020). Cost, Activation, Loss Function‖ Neural Network‖ Deep Learning. What are these? (online resource). Available on http://www.ai-shed .com[Date of access: 07/02/2021].

Sanderson Grant. (2017). What is backpropagation really doing? | Deep learning, chapter 3, 2017 (online resource). Available on http://www.ai-shed.com[Date of access: 07/02/2021].

9

Conclusions and Future Developments

A process is a set of interrelated or interacting activities which transforms inputs into outputs

[ISO 9000:2000]

In this book we have touched on various topics, some quite complex, especially for those unfamiliar with the physical–mathematical world. We believe that we have satisfied your curiosity about neural networks and fulfilled the promise made in the first chapter of providing the foundations as well as many interesting examples on how and in which circumstances neural networks can have a positive impact on our lives and work. Therefore, we would like to end with some observations and suggestions aimed at teasing your intellect.

We presented some examples of Neural Networks applied to sports clubs that, due to their organizational nature especially, adopt standard processes often based on experience.

A complex system, operating in an environment, would struggle to adapt and observe events autonomously without helping from the human side. The decision-making phase, together with an optimization process, brings the system to a next state that may or not may depend on its initial state or on previous states. Therefore, in all conditions the system deals with all the uncertainties that are proper to the surrounding reality.

All the information, obtained from the environment and appropriately combined with each other, are used to build an autonomous and intelligent system. Consequently, this System (even more complex) can generate a product or service necessary for specific requests or objectives. However, it is completely indisputable that an intelligent system would make the best decision through techniques related to its sensorial capacity and its deep learning ability.

Systems Engineering Neural Networks, First Edition. Alessandro Migliaccio and Giovanni Iannone.
© 2023 John Wiley & Sons, Inc. Published 2023 by John Wiley & Sons, Inc.

In the cases studied in this book we have emphasized that the human must help the system in measuring and evaluating events so that it can provide to human the best possible decision. We believe that, according to a system engineering approach, the relationship between human (system engineer) and system (integrated neural network) is two-way and continuously exchanged information.

The logic, behind the reasoning, has been widely used to understand reality, in fact the fuzzy method usage has been mentioned in some cases or Boolean in others. The logical arguments could be used to make a decision; however, they would not be enough to progress in the analysis of the event.

Consider, for example, the Figure 9.1, fascinating in some ways, but difficult to explain when observed through sensory stimuli, analyzed and interpreted through intuitive, psychic, and intellectual processes: each of us can give a personal interpretation. Here comes our Pythagoras to the rescue again, he explains us that, once we have identified the number, we can derive the structure of all reality-that is, in all things there is a mathematical regularity. To find a correspondence between reality and mathematics, we must go through empirical methods, but we have seen that, in the training phase, an exact reciprocity is required between the real solution of the problem and the calculated result. Could this approach be enough to state that neural networks are reliable?

When we talk about neural networks or about project optimization processes, or again if we want to use a mathematical approach to understand some dynamics of our daily life, we have to use different points of observations as one would surely not be enough.

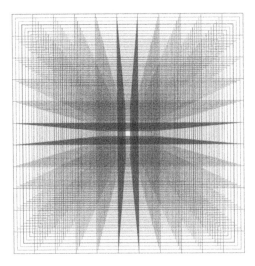

Figure 9.1 Can we understand reality purely through mathematical laws? The challenge of Artificial Intelligence. Source: Pixabay.

Through observation, we have critically investigated the possible variables of a system, affirming several times that it is difficult to find an absolute result even if it is a consequence of a great amount of information being processed.

It may happen, during the calculation phase, that we accept or neglect some variables because they are of a smaller order of magnitude than the others, or to transform some variables into constants to study aspects of the phenomenon in question.

Contradictory elements or biases could be met in all the processes assumed during the system engineering approach. All the information must acquire a certain value or lose it (weight). Hence, the plausible or the stochastic or the random reasoning is the fundamental management of a complex system using artificial intelligence.

The power of neural networks, or in general of the optimization of a problem, lies in the fact that we can consider many variables that describe first and then determine an event in all its complexity. At this point, we cannot help but accept that this method is reliable because it is of a comprehensive and adaptable nature.

We have implicitly stated that, to understand the solution to a problem or to make a prediction, we need to know the causes that determine it. These causes are to be found in the phenomenon itself by seeking its details and converting the data acquired from reality into mathematical symbols.

According to the Greek philosopher Aristotle, the logic (analitic) is defined as the science that studies the rules of scientific knowledge. We can reach true conclusions if the introductions to the problem (inputs) are true; therefore, it is strictly necessary to check and verify the inputs from which the investigation begins (organized collection of data stored). This procedure aims to know an event through correct definitions, and consistent reasoning. The generalization of results (ouput), obtained through partial collection of input data (abstraction), is first obtained through logical procedures that will later be validated and verified in reality (verification and validation).

The well-known philosopher and mathematician B. Russell stated that logical-mathematical relations are born through logical constructions, as reasoning and explanation. Going beyond the concept of metaphysics[42], that is to search for a reality behind phenomena, he reiterated that mathematics can be used as a science of relationships. Logic, like set theory, creates relationships between objects that have the same properties, then moves on to mathematical propositions and their relationships to explain and organize reality.

Mathematical logic is essentially the logic of relationships. Relations between things (objects, bodies, entities, systems, and so on) are the facts - empiricism refers to the experience of facts. Finally, a computational method expresses the link between relationships and experience (See Figure 9.2).

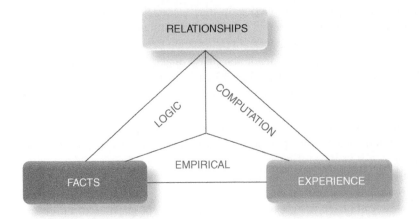

Figure 9.2 Reasoning Triangle- Continuous Process to reach a conclusion after thinking about all the facts and having gained a wealth of experience.

The more one understands these relationships, the more one can recognize them and different logical paths. **Rules** are useful to express the principles of cause and effect, after which the unknown variables of the problem could indicate other possible solutions. Will we be able to see other possible solutions and express ourselves without outside influence?

If each variable has a different effect on the **rule**, and so creates different results that are not predictable, then relying on the knowledge of causes and effects could be reductive. All potential choices not yet made are all still valid until we choose one that voids the others. Therefore, to make a change, one has to be above the rules, as knowledge takes us away from outer influence (bias).

Are axioms arbitrary or are they precise rules to be obeyed?

To be able to see all the possible options without being bound by the rules gives us the possibility to change them. If we change the rules, we change the initial axioms and we find ourselves faced with other possibilities. With logic, we can get to the wrong conclusions if starting from the wrong axioms. On the other hand, by observing contradictions, by relying on intuition or feelings, we can explore fields that go beyond reality. Our mind is bigger than the world that contains it. Any mathematical theorem that is proved cannot be complete and consistent at the same time. No theory is closed and complete, Kurt Godel said. It all depends on the initial axiom. It all depends on the choice of the axiom. We must immerse ourselves in reality to learn new rules and get out of the cause-and-effect stability, without transcending or going beyond.

Can we build machines that are able to autonomously elaborate multiple choices through a series of logical steps?

As we mentioned in the opening chapters, neural networks are only a part of the great world of artificial intelligence. Can we develop increasingly complex and advanced machines, and an increasingly inclusive relationship between machines and man? It would surely be interesting to try to transfer the concept of logic onto machines, as it characterizes the structure of thoughts and deductive reasoning. A machine able to autonomously analyze and validate facts, or even seek the truth of its own conclusions regardless of the content of the propositions involved.

A computational method does not guarantee absolute knowledge, which would be able to understand and process all that is possible to know. A computational method can only try to give us an exhaustive knowledge of the whole reality.

Therefore, tailoring methodology, especially applied when a Neural Network is introduced to increase the system autonomy or to optimize some system activities, is recommended. Despite the tailoring methodology are not supported by adequate guidelines, it has been seen that **rules**, based on a tailoring process and combined with project attributes, can be a viable approach to achieve the system objective, reducing the amount of efforts. Choosing and adapting robust methods is crucial to the application of system engineering principles. Therefore, we can use both top-down and bottom-up approaches for Neural Network tailoring process to increase the software efficiency.

In general, all the principles of artificial intelligence can be integrated into an existing system through a system engineering approach. Dividing the system into sub-systems and introducing them a neural network, any software functions perform an algorithm for the sub-system, or the sub-system is able to manage the integrated software (See Figure 9.3).

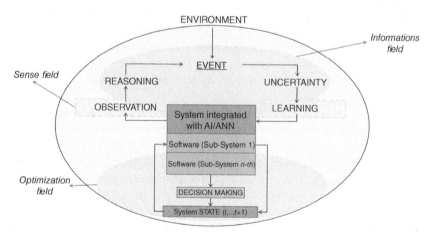

Figure 9.3 The contents of this book can be brought to its proper conclusions.

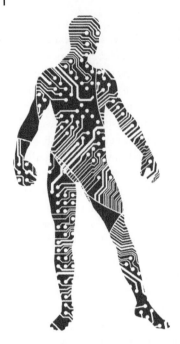

Figure 9.4 Can we develop increasingly complex and advanced machines, and an increasingly inclusive relationship between machines and man? Source: Pixabay.

Can we define a new concept of Man-Vitruvian by introducing artificial intelligence? In the Figure 9.4 we represent man according to the system and analytical-mathematical principles described in this book. Beginning with Descartes' (Cartesio) definition of the human brain, i.e., that connecting organ between the human body, considered a machine (*res extensa*) and the soul (*res cogitans*), defined earlier by Plato, we can only limit ourselves to defining artificial intelligence (and learning by neural networks). It is able to put together systems (abstract machines) and human (sensing machines) insofar as they are endowed with those typical human characteristics such as, for example, decision-making capabilities. That is, it is therefore, not only intelligence understood as the ability to compute or to know data obtained according to an abstraction process.

And so, does experience help?

The code we have used allowed us to associate each analyzed object not to a precise meaning but to a category. Based on this, we have been able to search for explanations by associating a number to the sensitive experience. Experience, therefore, differs from other searches for causes, because it allows us to know the nature of the object from a general and non-specific point of view - such as that of physics, chemistry, biology, psychology. Starting from a higher or more general level, we can throw ourselves in the direction of concrete aspects concreteness as in this process we will have laid the foundations for a deeper knowledge.

How can we transfer our knowledge to machines? Do machines need human beings to survive, or are they ready to exist without human help?

Knowledge has to be structured in such a way that all algorithms can operate optimally by providing correct outputs, that is, closer to reality. Machines associated with artificial intelligence can be used to give assistance in all those extreme working conditions human beings face, while they could focus on the management and control of processes, as well as on the machines themselves.

The use of collaborative machines (cobots) helps to tackle many challenges that are critical for human beings and susceptible to evaluation errors. Thanks to artificial intelligence, human beings are no longer responsible for interpreting, understanding, and analyzing data. Machines and human beings are surely compatible, as long as we are able to freely manage the outcome of a process initially regulated and subsequently carried out by a machine.

The search for knowledge on neural networks, and many other topics, continues on the blog associated with this book. On the blog, you will find many ideas and applications, and the opportunity to freely share your inspirations with the authors.

Glossary and Insights

In this section, you will find clarifications and insights on many concepts mentioned in the book. Additional sources will be indicated for further study. Given the dynamic nature of web links, online resources are featured in the blog associated with the book.

1	AlphaGO and AlphaFold	Have a look at the blog associated with the book – http://www.ai-shed.com
2	Reliability	The likelihood that a product, system, or process will perform its intended function for a specific period of time and operate in a defined environment without failure. The main characteristics of a reliable product are the reliability, the expected function, the degree of compliance, service life, and operating conditions (temperature, speed, pressure, and so on).
3	Machine learning	It is a branch of artificial intelligence (AI) that uses different methods for pattern recognition. It uses statistical methods to improve the performance of an algorithm in identifying data patterns. In computer science, machine learning is a variant of traditional programming, in which a machine can learn from the data independently and without explicit instructions.
4	Deep learning	It is a set of techniques based on artificial neural networks organized in different layers.

Source: Polytechnic University of Milan

Systems Engineering Neural Networks, First Edition. Alessandro Migliaccio and Giovanni Iannone.
© 2023 John Wiley & Sons, Inc. Published 2023 by John Wiley & Sons, Inc.

5	Supervised learning	It instructs a neural network to allow it to automatically make predictions based on a set of ideal examples (targets). For each input there is an output. This pair of values, initially provided (training set), allows the network to generate a function capable of obtaining the desired results for all the other non-provided samples (test set). Output values can be of a quantitative type (regression) or qualitative type (classification).
6	Binary classification	If we put an image into the network, it can be classified as belonging to one of two types: "cat" or "not." That is called two-class classification or binary classification.
7	Constructive feedback	Information that helps the receiver to identify some solutions in their areas of weakness. This information is provided with positive intentions and is used as a supportive communication tool to address specific problems or concerns. The aim is to give an individual the opportunity to improve or correct some of their shortcomings. One or more pieces of constructive information support the individual's personal and professional growth. Constructive feedback is the opposite of destructive feedback, which results in a direct attack on the individual.
8	Attention based convolutional neural network	In mathematics, convolution is associated with the principle according to which, given two functions, a third one can be generated that explains how the shape of the first one influences the second.

A convolutional neural network is often used to process images and examine their pixels. The convolutional process applies filters to the image to make it easy to process it without losing the main characteristics that are useful for prediction.

Attention, in machine learning, is similar in principle to the cognitive process in human beings, with the difference that the machine learning process only focuses on the important parts of the input data. |

| 9 | Derivative | In calculus, the derivative of a function in R is the limit tending to zero of the incremental ratios - the latter is defined as the ratio between the difference of the values assumed by the function for two values of the variable (x and $x + \Delta x$) and the variation of the values of the variable Δx. |

$$f'(x) = \frac{df(x)}{dx} = \lim_{\Delta x \to 0} \frac{f(x + \Delta x) - f(x)}{\Delta x}$$

It represents the rate of change of a function with respect to a variable, i.e. the measure of how much the growth of a function changes as its argument changes.
Geometrically, it is the slope of the line tangent to the curve that represents the function under consideration.

| 10 | Partial derivative | In calculus, the derivative of a function in R^n is the limit tending to zero of the incremental ratio defined as the ratio between the difference of the values assumed by the function for two values of n-th variable, for example x ($x,y,...$ and $x + \Delta x, \Delta y,...$) and the variation of the values of the n-th variable, for example Δx. |

$$f'_x(x, y, ...) = \frac{\partial f(x, y, ...)}{\partial x} = \lim_{\Delta x \to 0} \frac{f(x + \Delta x, \Delta y, ...) - f(x, y, ...)}{\Delta x}$$

| 11 | Structure fatigue | Mechanical phenomenon that leads to the breaking of a material subjected to variable loads over time, even if its elasticity limit has not been exceeded. A given material can break under fatigue even though the maximum intensity of the stress is below failure limit or yield limit. |
| 12 | Differentiable function | At a x_0 if there is a limit of the incremental ratio of the function as the increment Δx tends to zero (Figure G.1). A function that is differentiable at one point is a function that can be reduced linearly up to an infinitesimal amount around the value of that point. All partial derivatives calculated at that point exist and the limits of the incremental ratios calculated for each variable of the function are finite. Differentiability is the generalization of the concept of the derivative function. Differentiability is applied to variable vector functions and allows to identify a tangent hyperplane for each point of the function. |

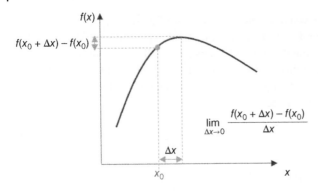

Figure G.1 Differentiable function.

| 13 | Gradient, slope | A vector whose components are the partial derivatives of a function with respect to the variables of the function itself. |
| 14 | Distributed computing | Distributed computing is a field of computer science that studies distributed systems, i.e. systems that consist of many autonomous interacting computers. These interact or communicate through a network to reach a common objective - a software running in a distributed system is called a distributed program, while distributed programming is the process of software writing. |

Source: Wikipedia

| 15 | Standard deviation | Given a population X of n data, it is possible to estimate how this population varies. It defines the variation of data around a position, in particular around the arithmetic mean. The standard deviation has the same unit as the observed values. |

$$dev.\ st = \sqrt{\frac{\sum_{i=1}^{n}(x_i - m)^2}{n}}$$

| 16 | Symplectic map | The definition of symplectic field is quite complex and it would require a separate section. If you are interested, we invite you to visit our blog to find articles related to this topic. |
| 17 | Surface corrosion | It indicates a natural and irreversible process of deterioration of the surface of a material. This phenomenon deteriorates the mechanical characteristics of the material. Its electrochemical nature determines a chemical–physical interaction of the surface of metallic material with the surrounding environment. |

18	Intergranular corrosion	It is a type of corrosion that occurs at the grain boundary of a metal in solid form. At high temperatures, and especially with steel, chromium carbides decrease in levels and diffuse toward the grain boundaries of the material, where there is greater concentration of carbon. The grain boundaries, rich in chromium carbides, act as a cathode, while the areas with a chromium percentage lower than 12% act as an anode: this is when wet corrosion can take place. A fine grain is counterproductive in this situation, as a greater extension of the grain implies a greater extension of the areas subjected to corrosion.
19	X-ray crystallography	It deals with the study of crystals, or crystalline grains or commonly grains (metallurgy) of a material. In particular, it studies the formation, growth, microscopic structure, macroscopic appearance, and physical properties of a material's grains.
20	Hamiltonian mechanics	This theory allows the probability and dynamics of a physical phenomenon to be reconciled through a formula. Based on the choice of new coordinates to generate the phase space, this theory rewrites the Euler-Lagrange motion equations and makes a scalar function match the total energy of the system. In general, in a dynamic system, the sum of potential and kinetic energy tends to zero.
21	Hornick's theorem	Or Universal Approximation theorem. Once we have a fixed activation function and positive integers, for each continuous function (target function) there is an approximate function (output layer) such that the calculated error is generally tending to zero. The theorem states that the result of the first layer can approximate any continuous function.
22	Human Factors	Often referred to as the application of psychological and physiological principles to human-operated processes and systems. The goal of human factors assessment is to reduce human error and improve safety during the operation during human machine interaction. The human factor is a multidisciplinary field of study, as it must examine human behavior to achieve occupational health, as well as safety and productivity goals. It must consider the capabilities and limitations of human beings when carrying out certain activities. In space and aeronautical fields, this discipline was developed to evaluate the adaptation between human beings and machines and between human beings and the environment.

23   Hyperplane   [...] generalization of the concept of plane, with which it coincides in the case of ordinary three-dimensional space. The word is in fact used to indicate a subspace of dimension $n - 1$ of a vectorial space (affine, Euclidean, projective) of dimension n. In a vectorial space Vn, a hyperplane is represented by a first-degree equation $a_1x_1 + ... + a_nx_n = 0$. In a Euclidean space, E_n hyperplane has the equation $a_1x_1 + ... + a_nx_n = k$, which can be rewritten using vector notation. Given $a = (a_1, ..., a_n)$ and $x = (x_1, ..., x_n)$, the hyperplane equation is the result of the scalar product

$$a\,x = k.$$

The two sets made of the solution points of the equations $a\,x \geq k$ and $a\,x \leq k$ are called hyperspaces of the original hyperplane $a\,x = k$. In particular, the hyperplanes of the Euclidean space E_3 are ordinary planes, while the hyperplanes of E_2 are straight lines.

Source: Treccani

24   Hyperspace   In mathematics, it is a multi-dimensional space. The number of these is usually indicated by n, in which case we also speak of space of dimension n - since ordinary space is three-dimensional, the hyperspace is a space of dimension $n > 3$. The notion of hyperspace may seem linked to artificial constructions, but it arises spontaneously from problems in which it is natural to consider variable entities depending on many parameters. N dimension Euclidean hyperspace: the points of this hyperspace are the *n-th* ordinates of real or complex numbers $(x_1, ..., x_n)$ - real or complex Euclidean hyperspace respectively - also called coordinates of the point itself. Its structure is obtained by translating the structure of metric space of the ordinary Euclidean space in three dimensions, by means of an analytical tool [...]

Source: Treccani

25   Separable space   In physical science [...] also known as locality of reference [...], according to which reality is made up of entities (particles, fields) located in several regions of space [...]

Source: Treccani

26 Event   In probability, one of the cases that can occur with certain
                probabilities (e.g. when tossing a coin, the event of one of
                the two sides showing). In physics, it is a phenomenon
                occurring in a small space and a short time and it can be
                characterized by a point in spacetime - four coordinates,
                three spatial and one temporal.

                Source: Treccani

27 Convex   A function is convex on an interval if and only if for each
   functions   pair of points on the graph there is a segment above or
                coinciding with a part of the graph itself.

28 Statistical Given a vector $\underline{v}$ of n components.
   mean (Chisini
   mean) $$\bar{v}(x, y, z, \ldots)$$

                and a function f of n variables, the mean is a real number m
                such that:

                $$f(x, y, z, \ldots) = f(m, m, m, \ldots, m)$$

                In fact, once the function is known, we can obtain different
                types of mean. For example, if the function is defined as the
                sum of its variables, then we have the arithmetic mean.

                $$f(x, y, z, \ldots) = x + y + z + \ldots$$
                $$x + y + z + \ldots = m + m + m + \ldots m = n \cdot m$$
                $$m = \frac{x + y + z + \ldots}{n}$$

                Similarly, if the function is defined as the sum of the square
                of its variables, we have the quadratic mean. We can
                continue to define the geometric and harmonic mean.

29 Computational [...] They are used to solve complex problems by means of
   methods   electronic computers. These complex problems can be
                formulated through mathematical language within various
                basic or applied science fields. Very often, these problems
                cannot be solved analytically since their solution does not
                allow for an explicit representation. We would be able to
                solve them analytically if the problems translated into
                nonlinear equations (algebraic, differential, or integral) of
                which we do not know the resolution formula. In other
                cases, the formula, even if known, is not usable to
                determine quantitative values of the solution itself, as it
                would require a prohibitive number of operations, even for
                the most modern electronic processors [...]

                Source: Treccani

30	Mathematical model	[...] scheme expressed in mathematical language and aimed at representing a phenomenon or a set of phenomena. The scheme can be built with one of the many concepts or mathematical theories, as well as a combination of both: algebraic or geometric structures; algebraic or differential equations, ordinary or partial derivative; finite difference; stochastic process; probability theory; game theory; systems theory, and so on. Regarding the phenomena it refers to, a mathematical model can have a descriptive function or aim at a description that is as satisfactory as possible, in order to allow a forecast about their future trend. This forecast can be limited to delineating this trend only in qualitative terms or determine it in exact quantitative terms - maybe through numerical computation assisted by an electronic computer. However, in relation to certain types of phenomena, the mathematical model can (or must) perform a prescriptive or control function to indicate how the phenomenon must take place to respond in the most effective way for a specific aim [...] Source: Treccani
31	Pixel	It is the unit that measures the surface of a digital image. The number of pixels arranged in a grid allows us to define the resolution of an image. The resolution, which establishes the level of detail in an image, is measured in pixels per inch (PPI). The higher the number of pixels per inch, the better the resolution. A higher resolution image usually results in a better print quality.
32	Prediction	The act of predicting, of assuming what will happen or how events will take place in the future, based on certain clues, or inductions, hypotheses, or conjectures. Source: Treccani
33	Forecast	Forecasting is a sub-discipline of prediction and consists in predicting future events based on precise data of present and past events.
34	Uncertainty principle	In a physical system, the uncertainty principle establishes the limits in measuring the values of physical quantities that are connected or incompatible with the system itself. $$\Delta x \cdot \Delta p_x \geq \frac{h}{2}$$ Where Δx is the uncertainty on the position of a particle and Δp is the uncertainty of the momentum of the same particle, h is the reduced Planck constant.

35	Primer	It is a protective coating (often in the form of paint) that is first applied to a generally metallic surface so that subsequent finishes can adhere.
36	Scalar multiplication	It links a real number to a pair of vectors - the number is calculated by adding the products between the homonymous components of the vectors.
37	Matrix multiplication	Also called row-by-column multiplication. Given two matrices, the scalar product is obtained between the first row of the first matrix and the first column of the second matrix, and so on. The result is a matrix having the number of rows equal to the rows of the first matrix and the number of columns equal to the columns of the second matrix.
38	Raspberry Pi	Microcomputer used for small projects and educational purposes. For more information, have a look at the blog associated with the book: http://www.ai-shed.com.
39	Convolutional neural network	It is a type of feed-forward artificial neural network in which the connectivity pattern between neurons is inspired to the organization of the animal's visual cortex, in which individual neurons are arranged in such a way to respond to the overlapping regions dowelling the visual field.
40	RGB	It is an additive color model that combines Red, Green and Blue - for example, Yellow is Red + Green, Magenta is Red + Blue, Cyan is Green + Blue. In addition, the sum of the three colors results in White and their total absence in Black. Due to its characteristics, it is a particularly suitable model for the representation and display of images in electronic devices. Most devices normally use combinations of Red, Green and Blue to display the pixels of an image, though this is particularly dependent on the device itself. The same image could be displayed in different ways if viewed on two different devices, as the material used to make the screens varies according to the manufacturer. On the other hand, this model is not suitable, due to the way in which it operates, for use in printers: image printing is performed by superimposing pigments, where each pigment reflects some luminous frequencies and absorbs or filters others. So, a typical printer will use a subtractive color model such as CMYK and a non-additive color model such as RGB. Vice versa, a juxtaposition of the pigments allows the realization of additive synthesis, e.g. screen coloring with offset between the layers, like the pictorial realization of pointillism.

41   Matrix   In mathematics and its applications, it is a rectangular table of symbols (called elements of the matrix). Usually these are representative of real or complex numbers, arranged by rows and by columns (rows and columns of the matrix) and indicated by a letter affected by two indexes (row index and column index) such that the element found in the m-th row and the n-th column is written a_{mn} [...]

Source: Treccani

42   Metaphysics   [...] according to Aristotle, it is the doctrine he called "first philosophy" [...] and defined as the theory of "being as being" [...], which studies the reality considered only in those very universal characters that make it so and excludes those specific characteristics that give it the nature of a determined reality, the object of a particular science. It is intrinsic, as we make this distinction, that a theoretical knowledge of reality has the character of absolute knowledge, with respect to the relativity of all others. In the history of thought, metaphysics sometimes presents itself as ontology, in realistic or objectivism philosophical systems. Other times, it identifies with psychology or gnoseology, with logic or dialectics, or even with ethics, in idealistic and subjectivism philosophical systems. As a general term, it is the denomination of any doctrine which presents itself as fundamental with respect to the sciences of relative and particular realities, and it defines itself as the science of absolute reality. Overall understanding of reality, of the world, of human life, not necessarily built on philosophical basis [...] General theory based at the foundation of science, or even art, technique, and treatise.

Source: Treccani

43   Normalization   The process of dividing all the terms of an equation by the same factor so that the resulting equation has a norm equal to 1.

44   Pattern   [...] word used to designate the scheme of an integrated circuit, or the configuration assumed by particular experimental results, and so on [...] From English: model, scheme, configuration.

Source: Treccani

45	Recursive neural network	It is the generalization of a recurring network. A recursive network is characterized by a matrix of weights, unlike the recurring network. Each row of the weight matrix (vector) corresponds to a single input vector. The network error is calculated for each input network.
46	Quasi-static process	Thermodynamic transformation [...]: process in which a system goes through a succession of statuses of balance, that is every instant in which thermodynamic quantities (temperature, pressure, and so on) have a determined and measurable value [...] Source: Treccani
47	Linear regression	A Linear function of an independent variable can be associated with a population of data. In statistics, regression analysis is associated with the resolution of a linear model.
48	Non-linear regression	A nonlinear function of an independent variable can be associated with a population of data. Unlike linear regression, there is no general method for determining the best interpolation for the data. We use classes of optimisation algorithms which, starting from initial values chosen at random or through preliminary analysis, reach optimal points. Unlike linear regression, we could have local maxima where the maximum is global in nature.
49	Fretting	It refers to wear and corrosion damage on contact surfaces. The damage is caused by the presence of repeated relative surface movements, often caused by vibrations or in general by a certain load. Fretting tangibly degrades the quality of the surface layer by producing greater surface roughness and micro pits, which reduce the fatigue resistance of the components. Contact movement causes mechanical wear and material transfer to the surface, often followed by oxidation of both metal debris and newly exposed metal surfaces. Since oxidized debris is usually much harder than the surface it comes from, it often acts as an abrasive agent that increases the rubbing rate.

50	Learning rate	It is a parameter that guides the optimisation algorithm toward a minimum or a maximum. It represents the speed with which the cost function manages to reach convergence. The learning rate is characterized by two factors: the value that defines the overshooting and the direction toward the minimum, this generally being the slope of the loss function. A too high learning rate would not allow for a precise calculation of the function minima, while a too low learning rate could lead to an undesirable local minimum. Therefore, to obtain a correct convergence, the learning rate is made to vary during the training phase in which techniques are introduced to adapt it to each step.
51	Somatostatin	It is a hormone produced by the hypothalamus in the pancreas and gastrointestinal tract. It inhibits the secretion of the growth hormone and the release of insulin, glucagon, and hydrochloric acid in the stomach. It acts as a neurotransmitter and has a stimulating action on cholinergic and β-adrenergic receptors. Its levels increase with age, and this leads to a decreasing of the growth hormone levels and greater risk of obesity.
52	Test set	It is a data matrix composed of as many n rows as there are samples, and as many m columns as there are variables or inputs of the neural network. This data sample is used to define the reliability of the network on data acquired also at a later time.
53	Training set	It is a data matrix composed of as many n rows as there are samples, and as many m columns as there are variables or inputs of the neural network. Group of input data and solutions, normally used for training the neural network.
54	Validation set	It is a data matrix composed of as many n rows as there are samples, and as many m columns as there are variables or inputs of the neural network. This data sample is used to validate the network or to fix the weights and bias values calculated through the training set.
55	Vector in R	N-th ordinate of elements where each element is a real number. Each element is called a component of the vector.

Index

Systems Engineering Neural Networks, First Edition. Alessandro Migliaccio and Giovanni Iannone.
© 2023 John Wiley & Sons, Inc. Published 2023 by John Wiley & Sons, Inc.

Printed and bound by CPI Group (UK) Ltd, Croydon, CR0 4YY

17/01/2023

03181432-0002